LeMoyne's Botanical Watercolours

A Book of Days

British Museum

Pomegranate Calendars and Books
San Francisco

Published by Pomegranate Calendars and Books
Box 808022, Petaluma, CA 94975

ISBN 0-87654-602-5
Pomegranate Catalog Number A561

For more information on Pomegranate's books of days and other publications (including calendars, address books, art books, print portfolios, books of postcards, notecards, posters and postcards), please write to Pomegranate, Box 808022, Petaluma, CA 94975.

Designed by Stephen Kruse
Printed in Singapore

Introduction

Jacques le Moyne de Morgues (*c.* 1533-1588) was a native of Dieppe. The first thirty years of his life are entirely undocumented, but it is probable that he received his artistic training locally, for Dieppe was a renowned centre for fine cartography and illumination. In 1564 he was commissioned to accompany a Huguenot expedition to Florida in order to carry out essential mapping duties and to make a graphic record of all he saw, notably the native Indians and the flora. In September 1565, the strongpoint established by the French, called Fort Caroline (near present-day Jacksonville), was attacked by the Spaniards, and most of the colonists were massacred. Le Moyne was one of the few who escaped, returning to France after enduring great privations. He later wrote his own account of the French expedition, and some of his illustrations of Indian life were posthumously published by Theodor de Bry in 1591. Le Moyne, a Huguenot, finally settled in England 'for religion', probably around 1580. In London his most important patron was Sir Walter Raleigh, and he also knew the artist John White (*fl.* 1585-93), who was to accompany Raleigh's 1585 expedition to Virginia. Both before and after the expedition to Florida, Le Moyne seems to have been occupied in making studies of plants. The watercolours illustrated in this book of days must have been painted within about three years of this death, since the prefatory sonnet bears his name and the date 1585. They are mostly familiar garden flowers and fruits, studied with a delicacy and distinction remarkable for the period. In 1586 Le Moyne published a series of woodcuts of plants (many of which were simplified versions of the watercolours now in the British Museum) and animals, dedicated to his patroness Lady Mary Sidney. Part bestiary, part florilegium – the earliest printed work of this kind to be produced in England – *La Clef des Champs* was intended as a pattern book for the embroiderer, jeweller or painter, and was Le Moyne's last traceable work.

Opposite page: Le Moyne's *Sonnet* was inscribed in the original French in pen and dark red ink, in italic hand, the alternate letters of the heading and the initial letters of each line picked out in gold. Later inscriptions on the lower half of the sheet give evidence of 18th century ownership: *Jòs Offley Esqr Ex Dono/Edmundi Bohun/Test/Carr Rich/Hannah Clues*. The handwriting has been identified as that of John de Beauchesne, the Huguenot writing-master.

21.3 x 14.2 cm (8 3/8 x 5 5/8 in.)
1962-7-14-1(1)

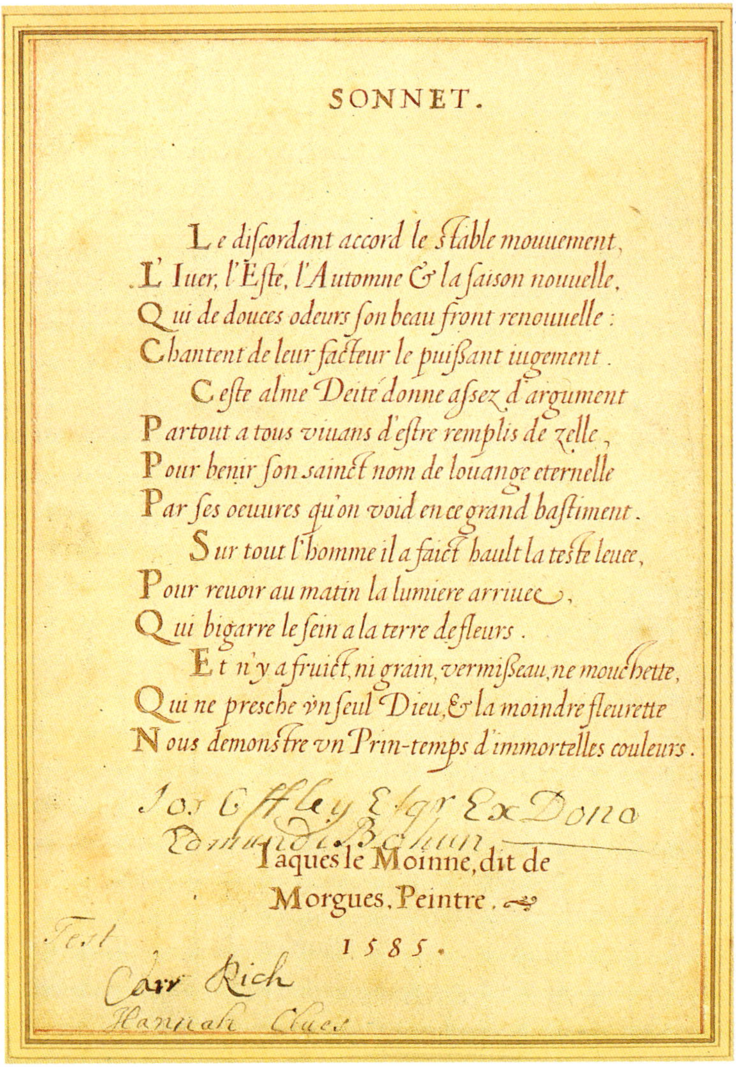

SONNET.

Le diſcordant accord le ſtable mouuement,
L'Iuer, l'Eſté, l'Automne & la ſaiſon nouuelle,
Qui de douces odeurs ſon beau front renouuelle:
Chantent de leur facteur le puiſſant iugement.
Ceſte alme Deité donne aſſez d'argument
Partout a tous viuans d'eſtre remplis de zelle,
Pour benir ſon ſainct nom de louange eternelle
Par ſes oeuures qu'on void en ce grand baſtiment.
Sur tout l'homme il a faict hault la teſte leuee,
Pour reuoir au matin la lumiere arriuee,
Qui bigarre le ſein a la terre de fleurs.
Et n'y a fruict, ni grain, vermiſſeau, ne mouchette,
Qui ne preſche vn ſeul Dieu, & la moindre fleurette
Nous demonſtre vn Prin-temps d'immortelles couleurs.

Jos: Offley Esqr Ex Dono Edmundi Bohun

Iaques le Moinne, dit de
Morgues, Peintre.

1585.

Test
Carr Rich
Hannah Clues

Sonnet

Discordant harmony and balanced movement,
Winter and Summer, Autumn, reborn Spring,
Renewing her sweet scents and colouring,
Join in the praise of God's unfailing judgment.

This loving God gives every argument
To look for zeal from each created thing,
To bless His Name eternally and sing
All He has made in earth and firmament.

Above all He made man with head held high
To watch each morning as new light arrives
And decorates earth's breast with varied flowers.

There is no fruit, or grain, or grub, or fly
That does not preach one God, the least flower gives
Pledge of a Spring with everlasting colours.

Jacques Le Moyne, called De Morgues, 1585

Translated from the French by R.N. Currey

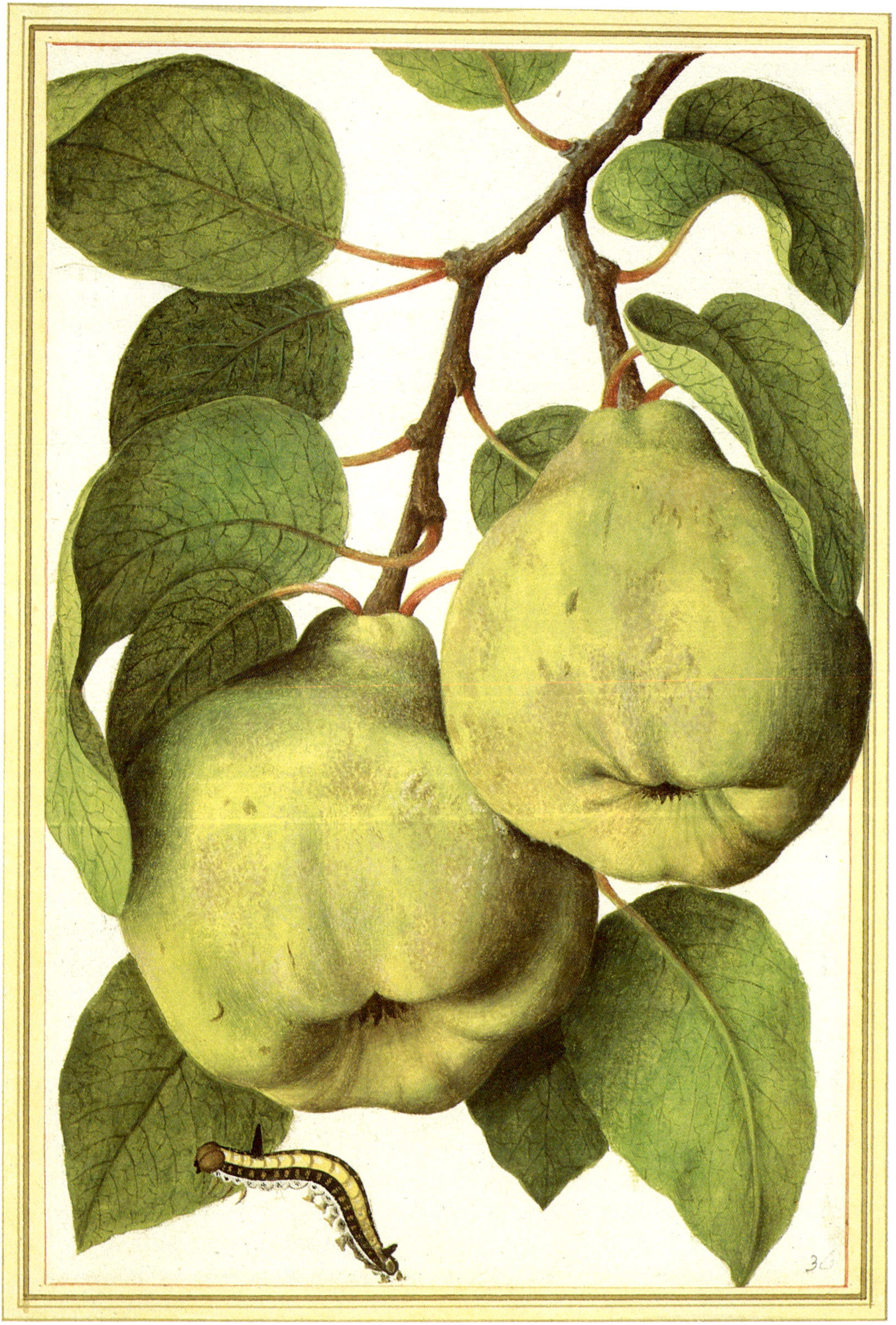

January

1

2

3

4

5

6

7

QUINCE AND CATERPILLAR
Quince, *Cydonia oblonga* Miller; Caterpillar of the Grey Dagger Moth, *Acronicta psi* (L.)
Watercolours and bodycolours with some oxidization on the caterpillar near the bottom edge; within a ruled red ink border, and extending (2mm) at one point beyond it on the right.
21.6 x 13.9 cm (8 1/4 x 5 1/2 in.) 1962-7-14-1(38)

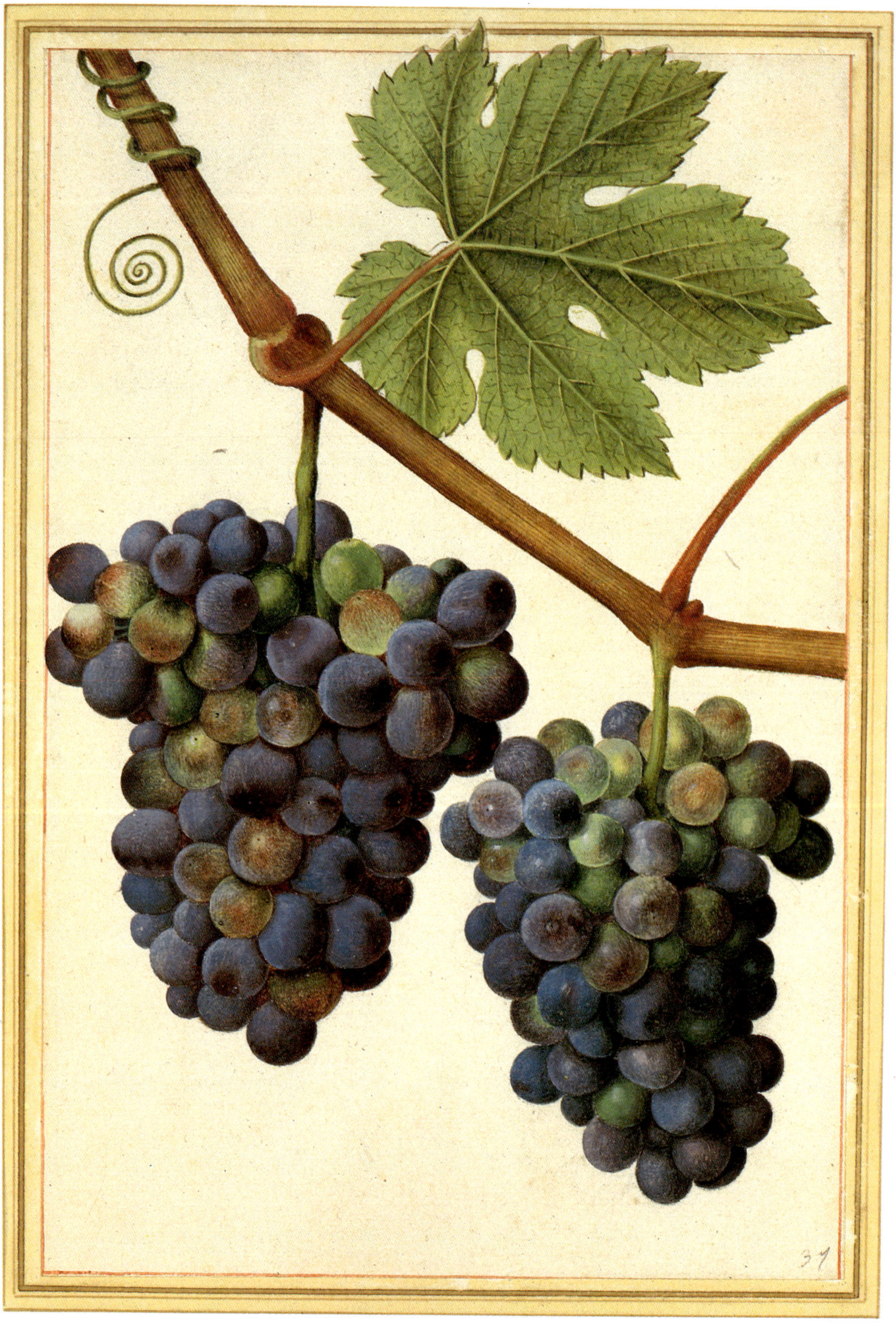
37

January

8

9

10

11

12

13

14

GRAPEVINE
Grapevine, *Vitus vinifera* (L.)
Watercolours and bodycolours within a ruled red ink border, and extending beyond it at the top (from 2 to 6 mm).
21.3 x 13.9 cm (8 3/8 x 5 1/2 in.) 1962-7-14-1(39)

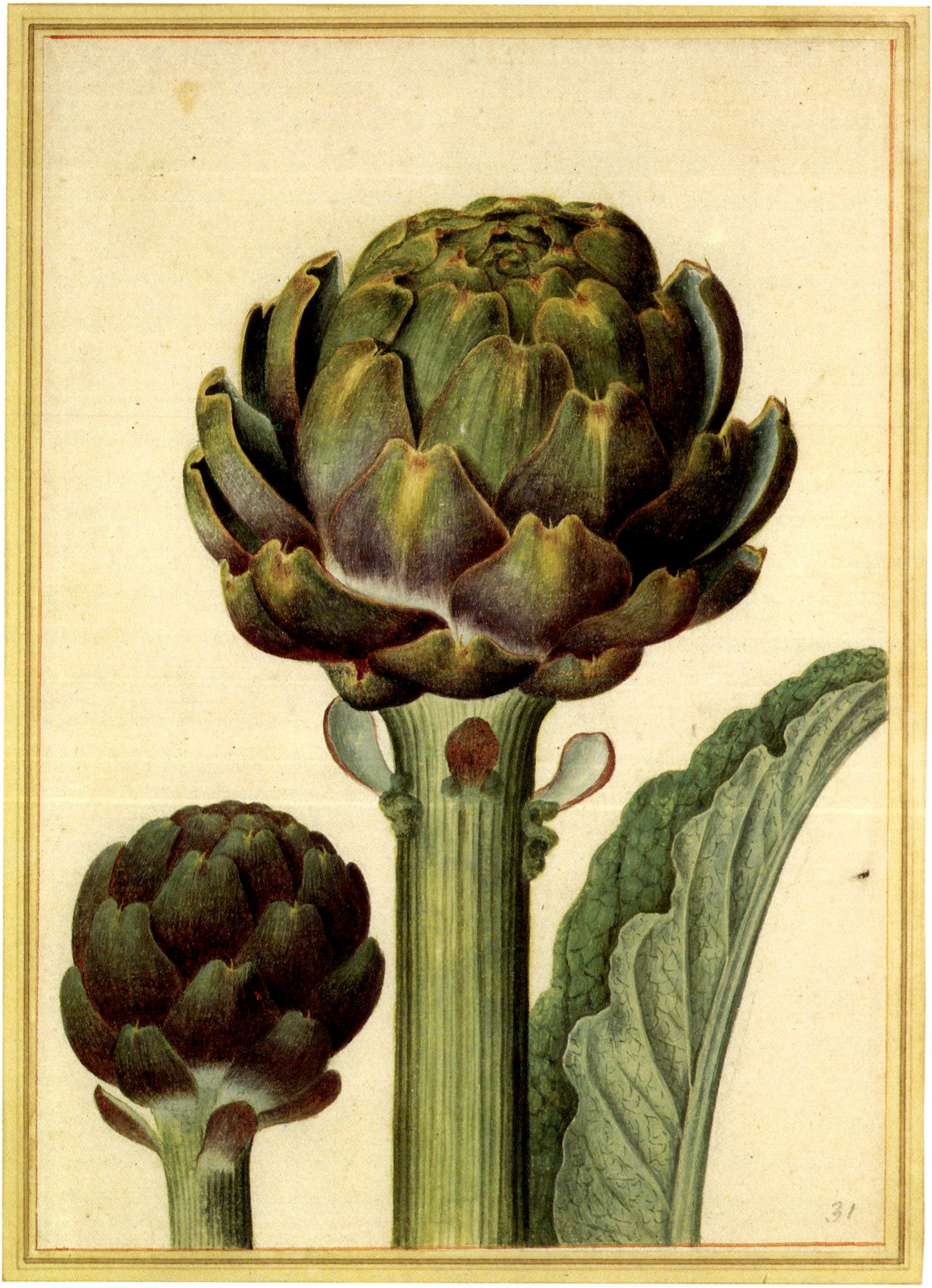
31

January

15

16

17

18

19

20

21

GLOBE ARTICHOKE
Globe Artichoke, *Cynara scolymus* (L.)
Watercolours and bodycolours within a ruled red ink border, partly cut away at the top.
20.6 x 14.5 cm (8 1/8 x 5 5/8 in.) 1962-7-14-1(33)

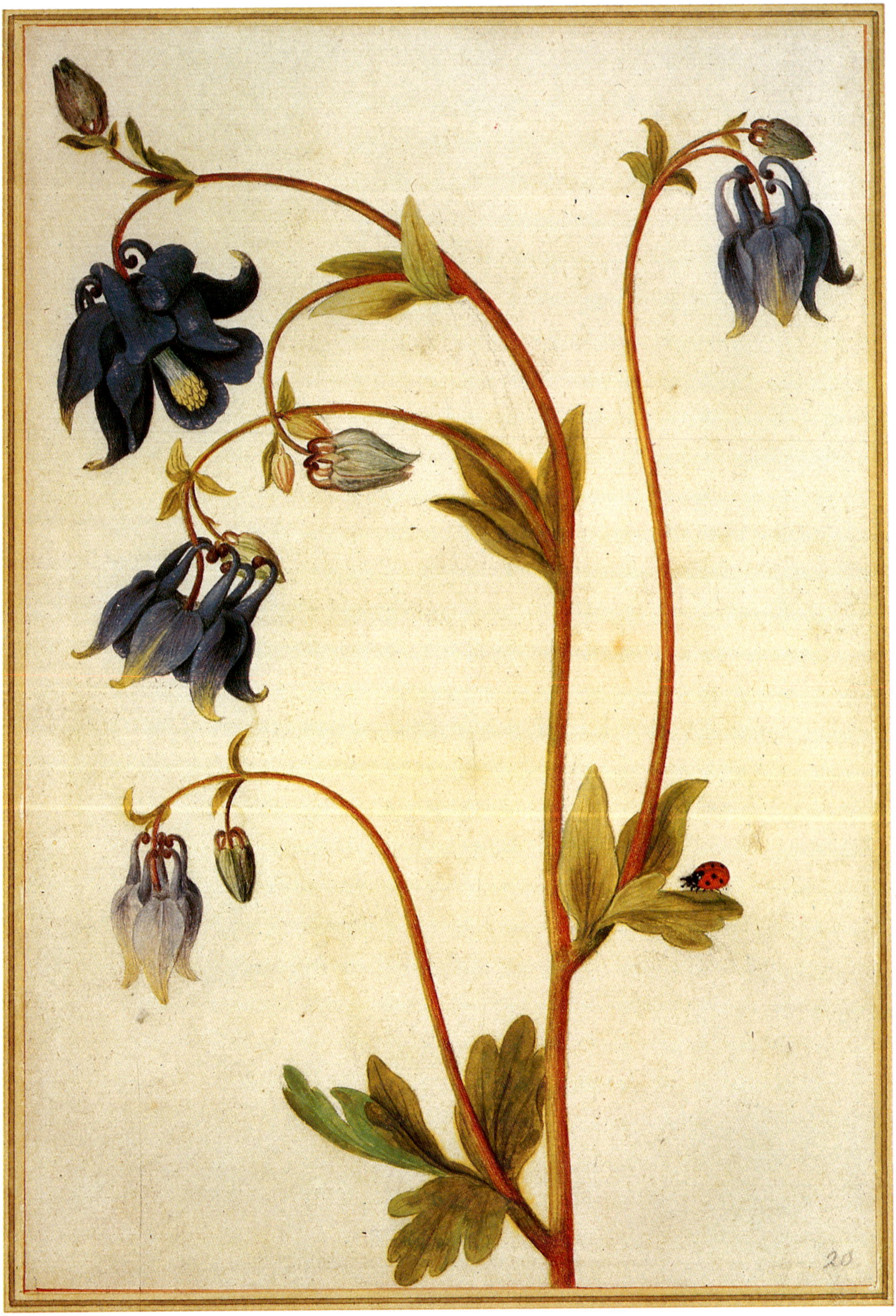
20

January

22

23

24

25

26

27

28

Columbine and Ladybird
Aquilegia (Columbine), *Aquilegia vulgaris* (L.); Ladybird, *Coccinellidae* family
Watercolours and bodycolours with very slight flaking on the two largest flowers; within a ruled red ink border, partly cut away at the top and right.
21.3 x 14.4 cm (8 3/8 x 5 5/8 in.) 1962-7-14-1(22)

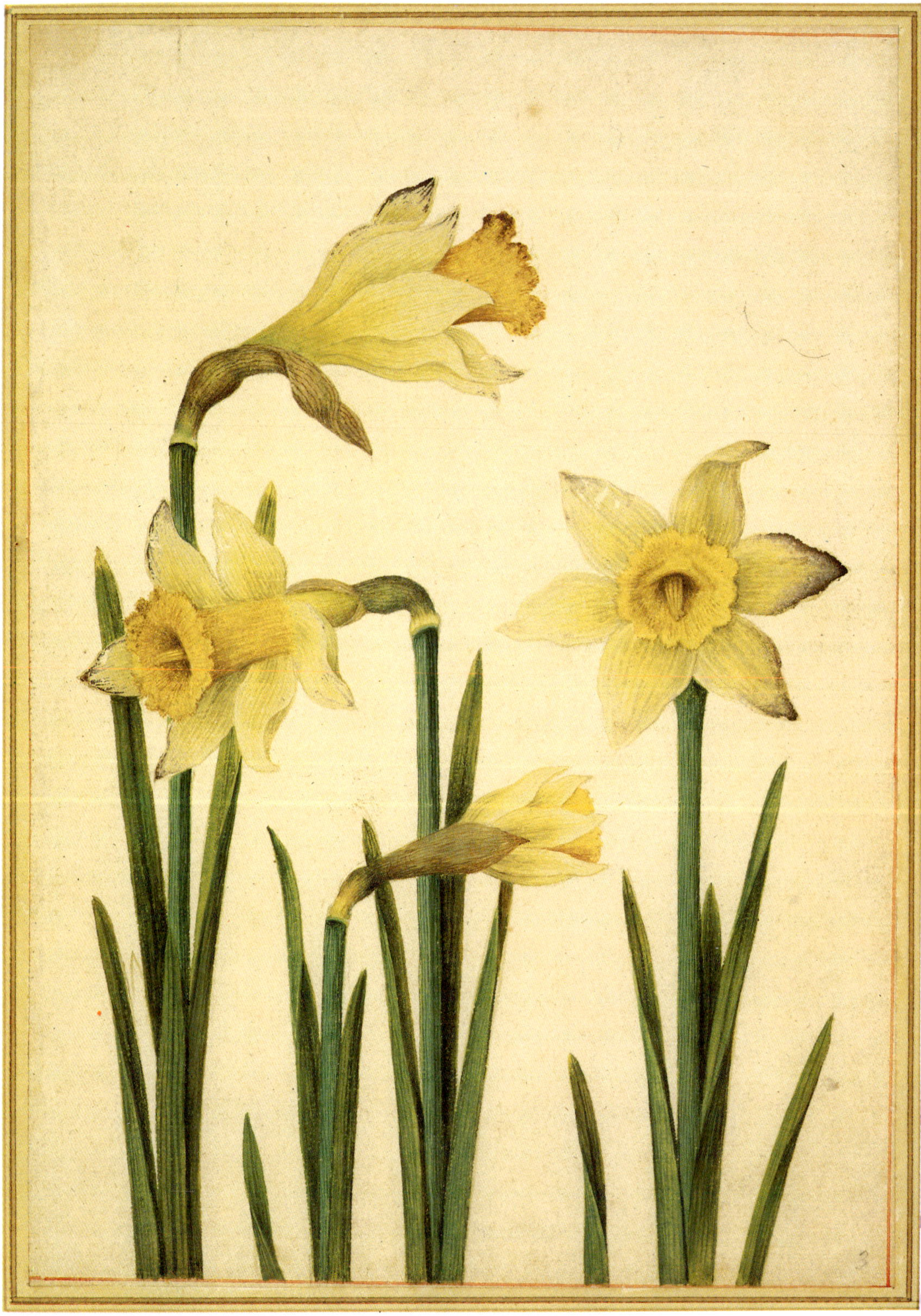

January/February

29

30

31

1

2

3

4

DAFFODIL
Wild Daffodil, *Narcissus pseudonarcissus* (L.)
Watercolours and bodycolours with oxidization near the edges; within a ruled red ink border, partly cut away at the top left and right.
21.7 x 14.8 cm (8 1/2 x 5 3/4 in.) 1962-7-14-1(5)

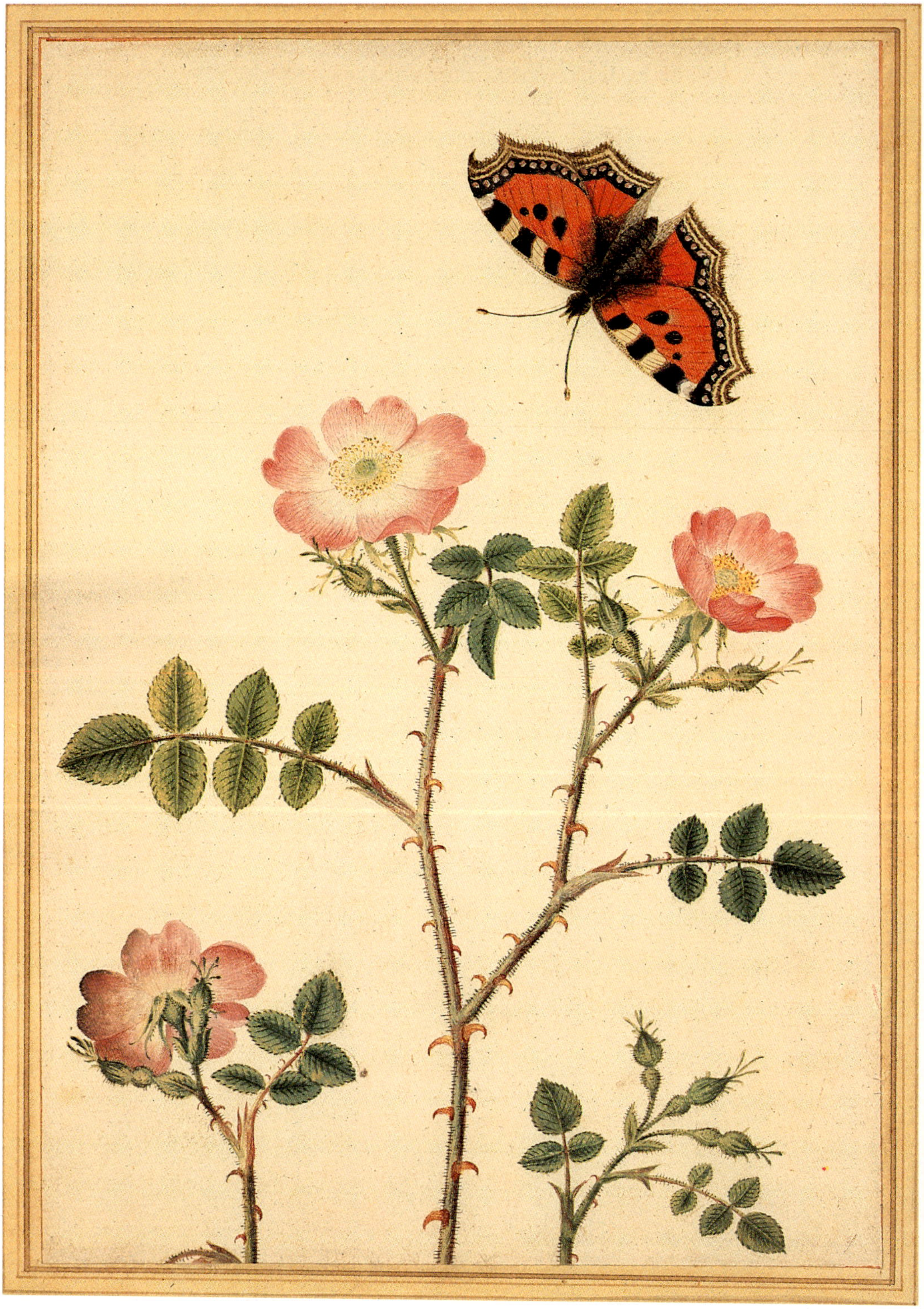

February

5

6

7

8

9

10

11

DOG ROSE AND SMALL TORTOISESHELL BUTTERFLY
Dog Rose, *Rosa canina* (L.); Small Tortoiseshell Butterfly, *Aglais urticae* (L.)
Watercolours and bodycolours with some oxidization near the lower left-hand border, within a ruled red ink border, partly cut away on all sides.
21.1 x 14.4 cm (8 1/4 x 5 5/8 in.) 1962-7-14-1(11)

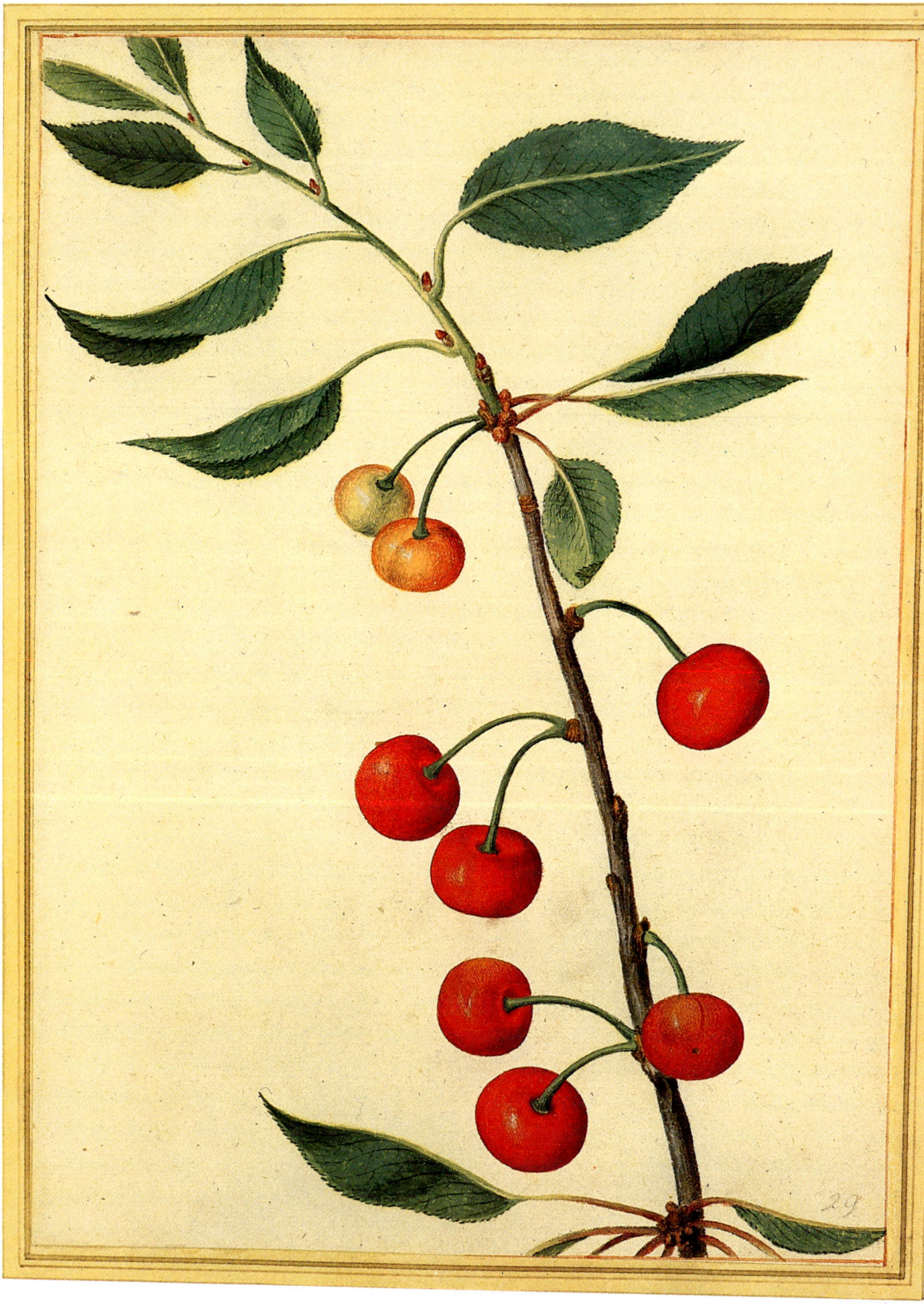
29

February

12

13

14

15

16

17

18

CHERRY
Wild Cherry, *Prunus avium* (L.)
Watercolours and bodycolours within a ruled red ink border, largely cut away at the left and bottom.
21.2 x 14.7 cm ($8\frac{3}{8}$ x $5\frac{3}{4}$ in.) 1962-7-14-1(31)

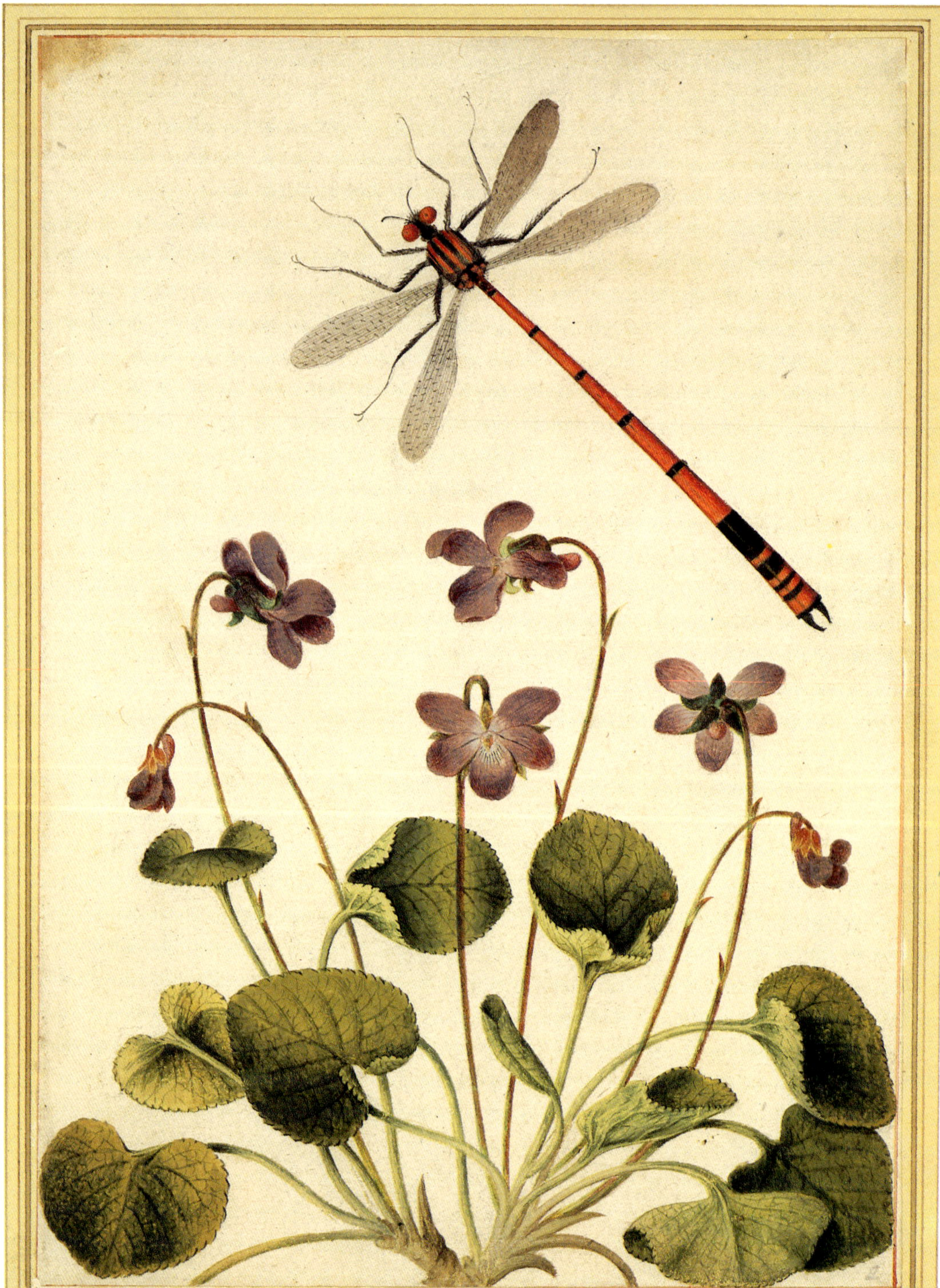

February

19

20

21

22

23

24

25

Dog Violet and Damselfly
Dog Violet, *Viola canina* (L.); Damselfly, *Pyrrhosoma nymphula* (Sulzer).
Watercolours and bodycolours within a ruled red ink border, partly cut away on all sides.
20.6 x 14.4 cm (8 1/8 x 5 5/8 in.) 1962-7-14-1(3)

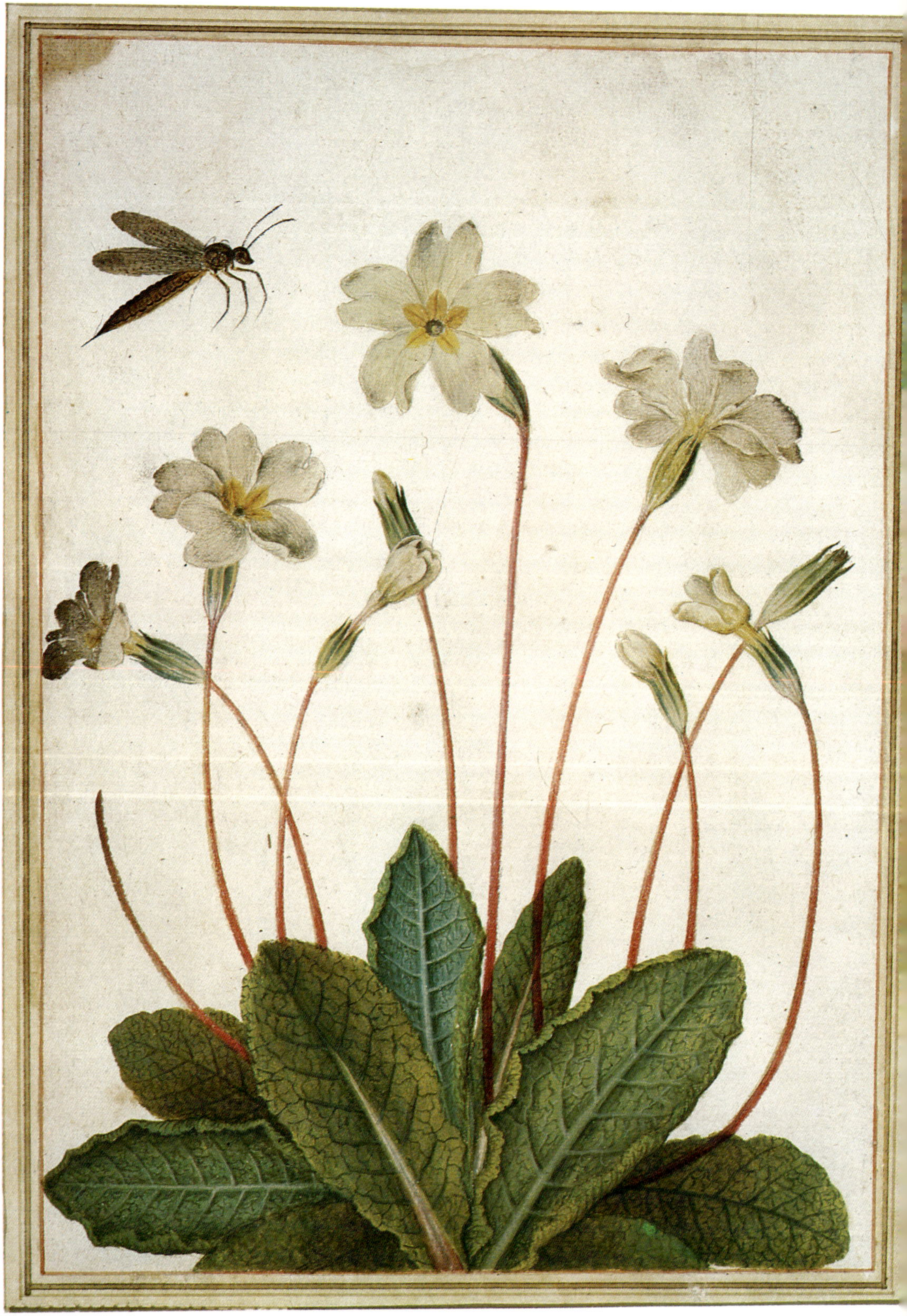

February/March

26

27

28

29

1

2

3

4

PRIMROSE AND FLY
Primrose, *Primula vulgaris* Hudson; Nematocerous fly
Watercolours and bodycolours with oxidization on the petal edges; within a ruled red ink border.
20.7 x 14.1 cm (8 1/8 x 5 1/2 in.) 1962-7-14-1(4)

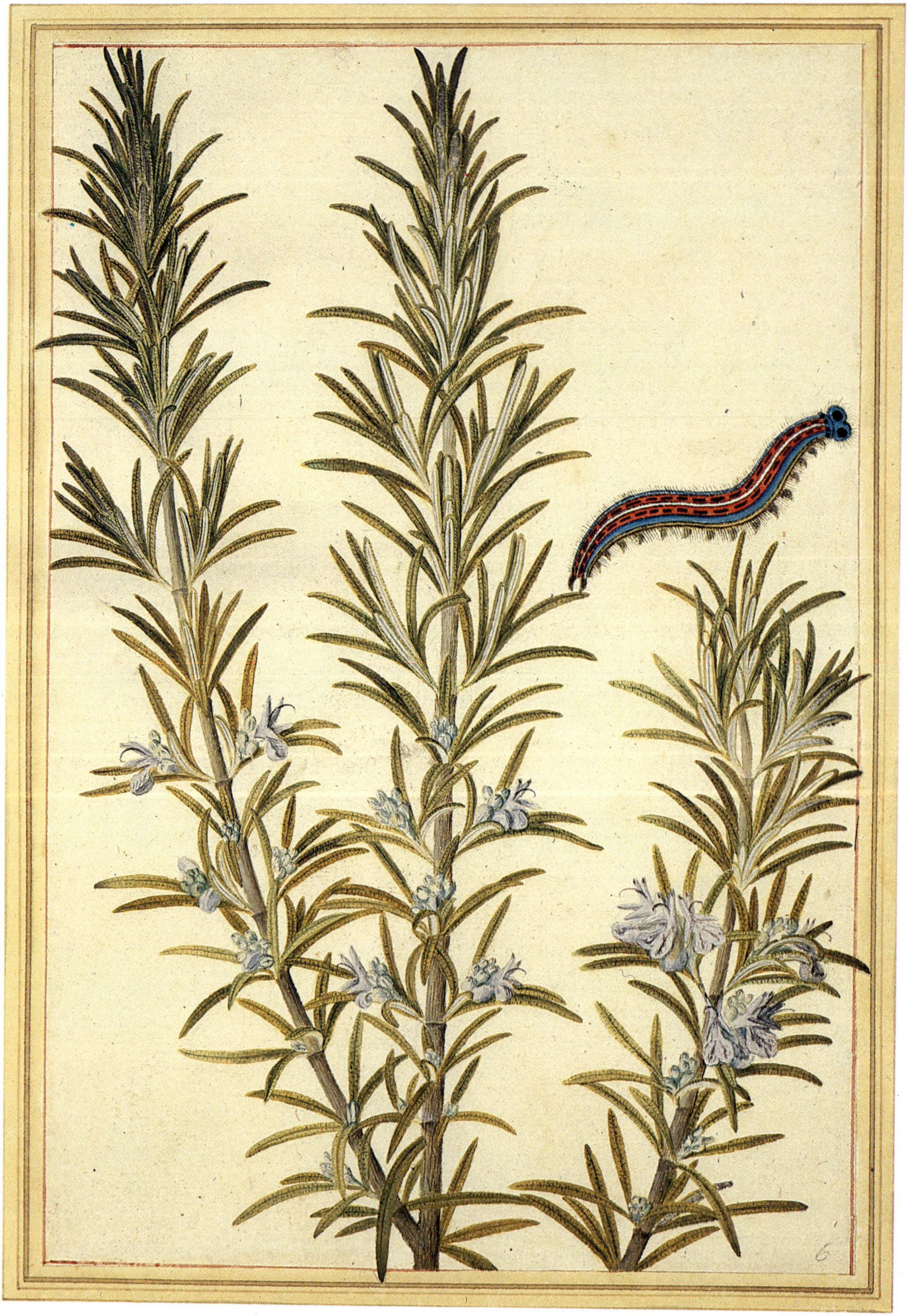

March

5

6

7

8

9

10

11

ROSEMARY AND LACKEY MOTH CATERPILLAR
Rosemary, *Rosmarinus officinalis* (L.); Caterpillar of the Lackey Moth, *Malacasoma neustria* (L.)
Watercolours and bodycolours within a ruled red ink border.
21.5 x 14.2 cm (8 1/2 x 5 5/8 in.) 1962-7-14-1(8)

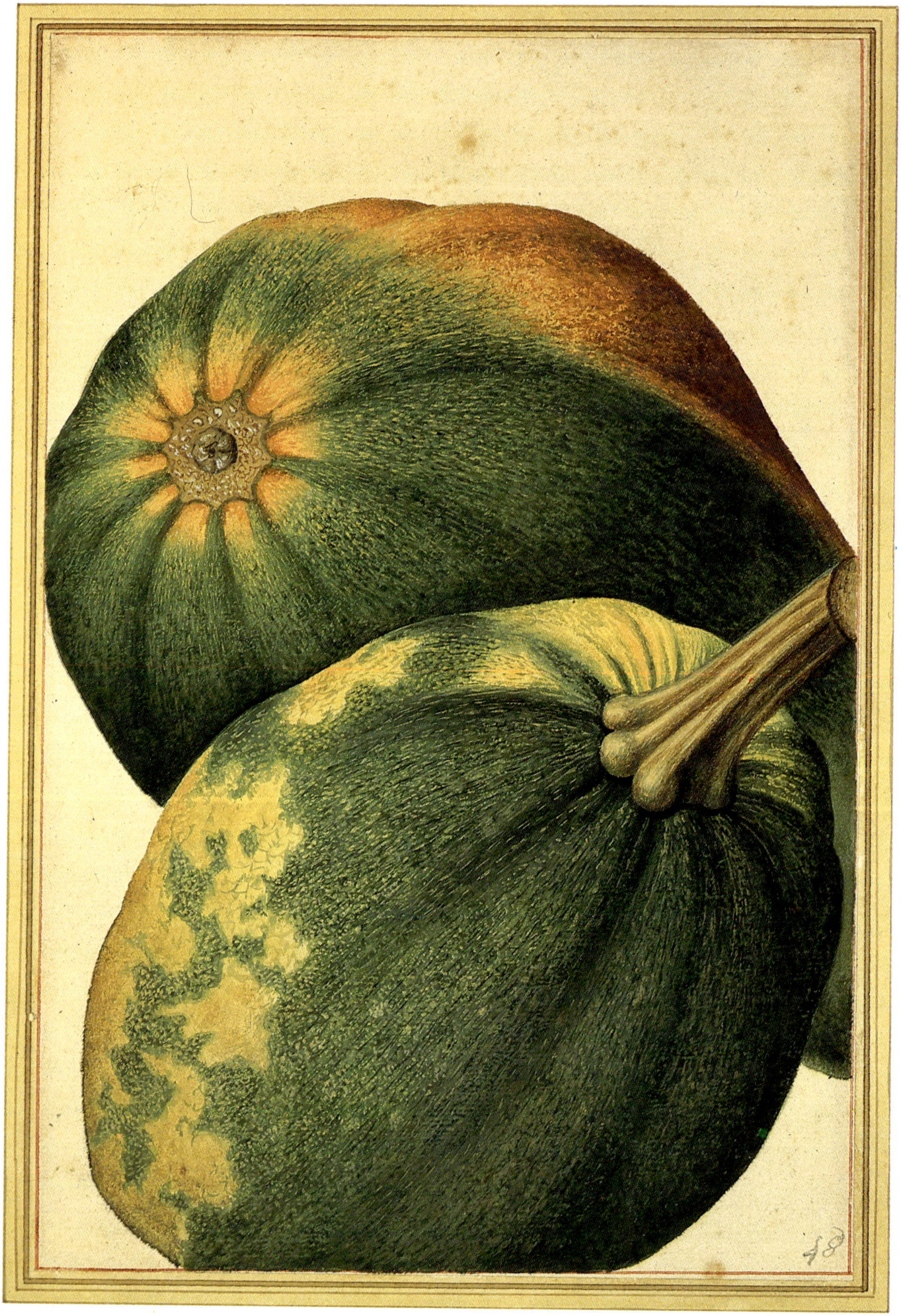

March

12

13

14

15

16

17

18

MARROW
Vegetable Marrow, *Cucurbita pepo* (L.), var. *medullosa* Alefeld
Watercolours and bodycolours within a ruled red ink border.
21.5 x 14.1 cm (8 1/2 x 5 1/2 in.) 1962-7-14-1(50)

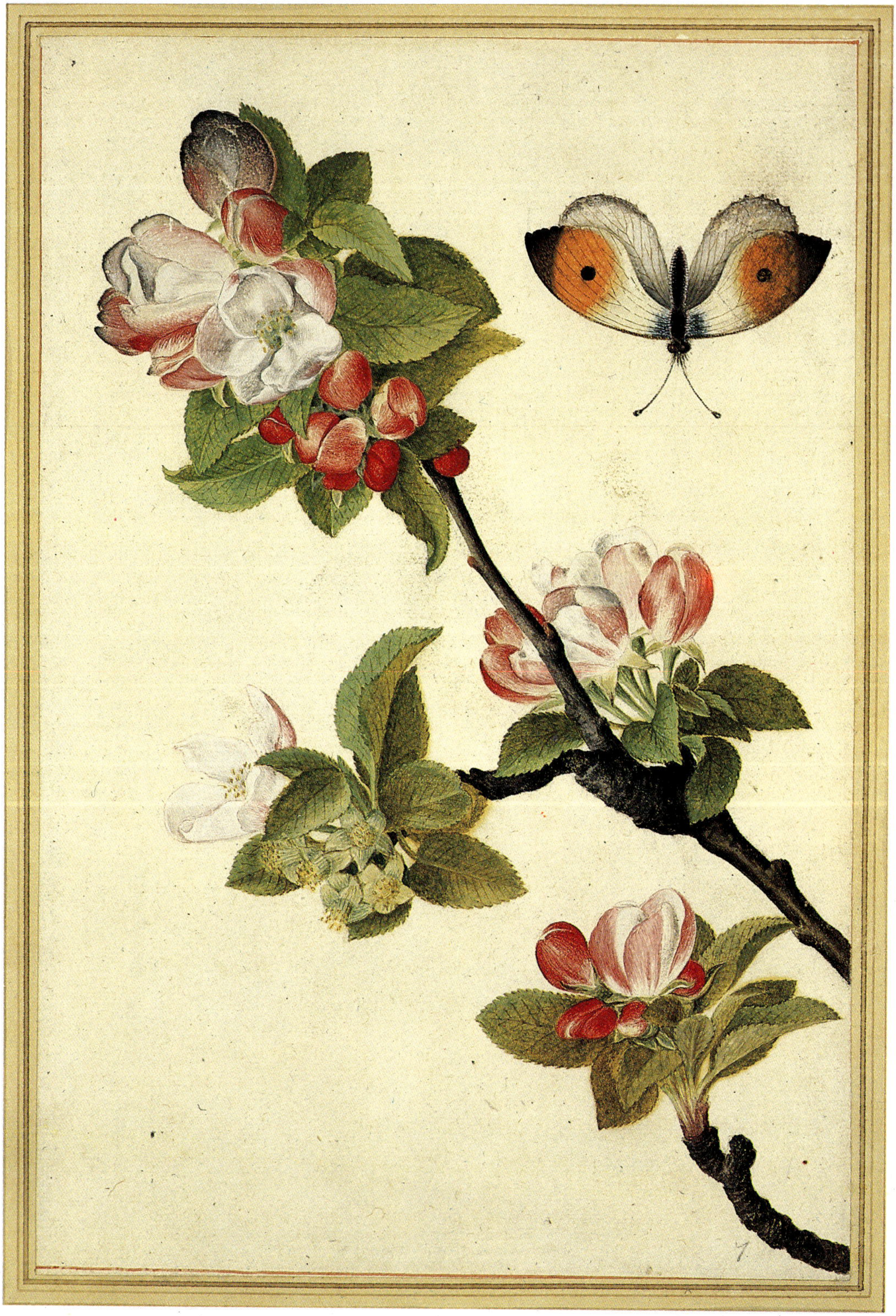

March

19

20

21

22

23

24

25

APPLE BLOSSOM AND ORANGE-TIP BUTTERFLY
Apple, *Malus pumila* Miller; Male Orange-tip Butterfly, *Anthocharis cardamines* (L.)
Watercolours and bodycolours with oxidization nearest the borders, particularly on rear wings of the butterfly, within a ruled red ink border, partly cut away on all sides.
21.7 x 14.2 cm (8 1/2 x 5 5/8 in.) 1962-7-14-1(9)

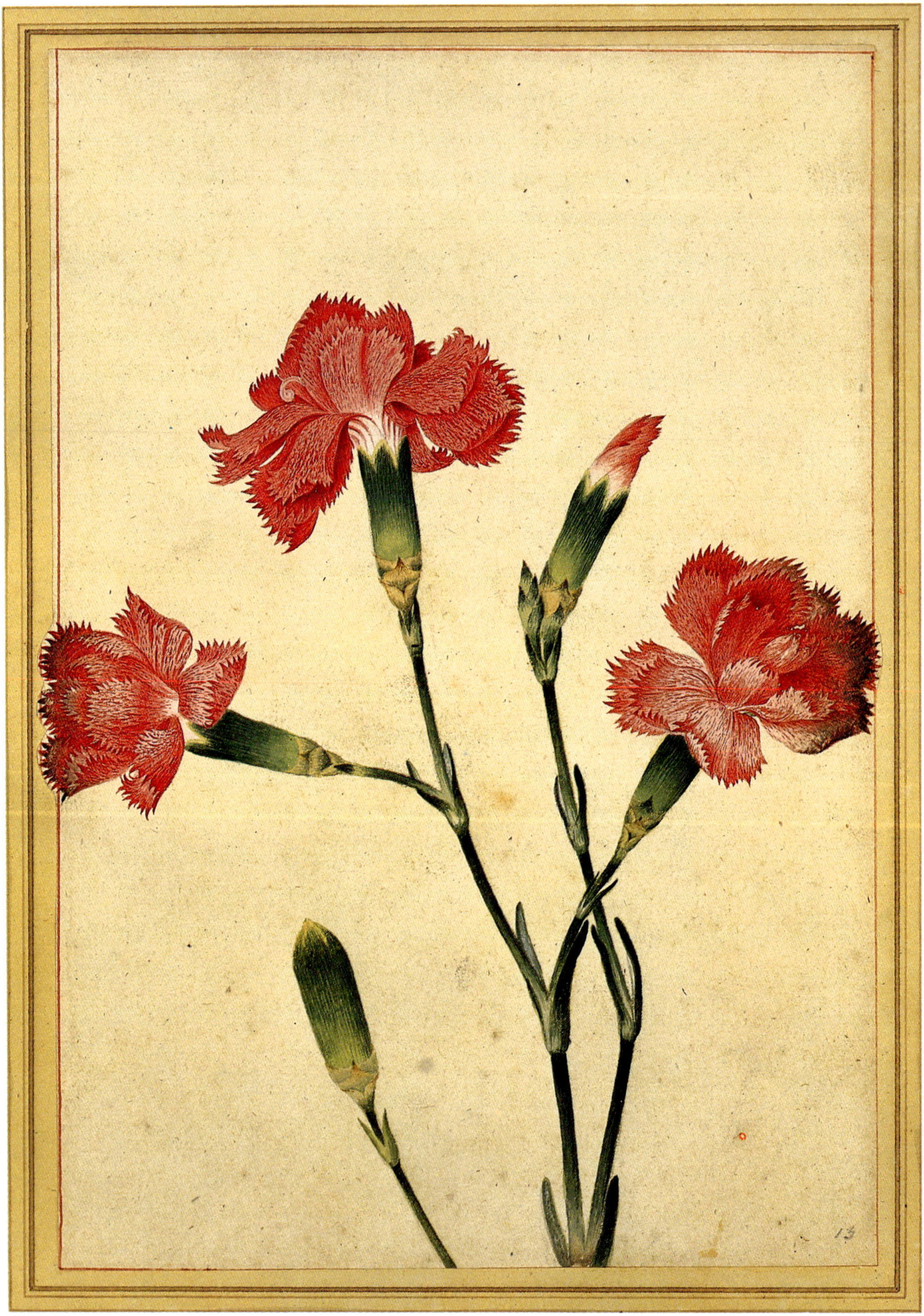

March/April

26

27

28

29

30

31

1

CLOVE PINK

Carnation (Clove Pink), *Dianthus caryophyllus* (L.) cultivar

Watercolours and bodycolours within a ruled red ink border, partly cut away at the bottom but with the flowers extending beyond it at the left and right.

21.2 x 14.3 cm ($8\frac{3}{8}$ x $5\frac{5}{8}$ in.) 1962-7-14-1(15)

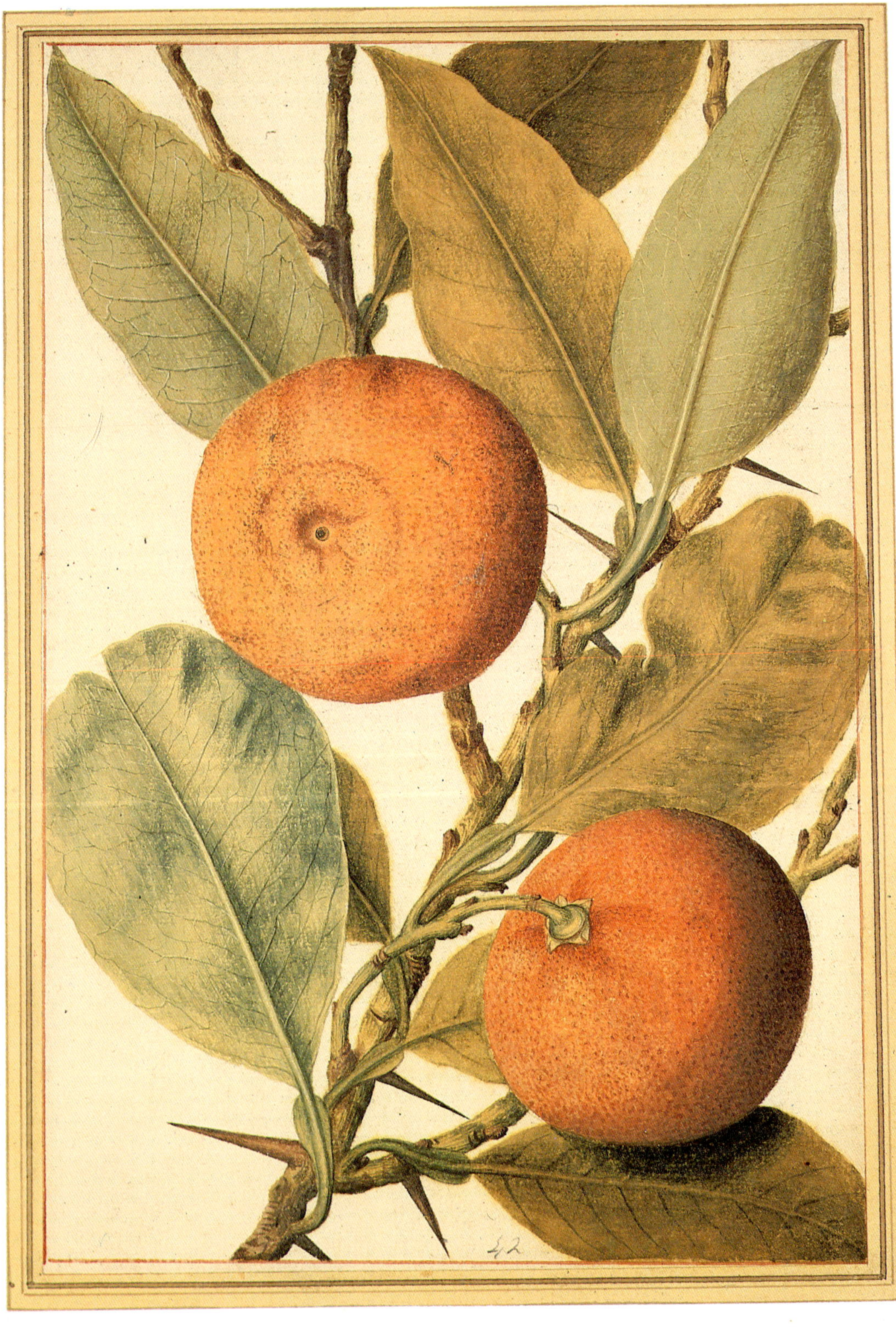

April

2

3

4

5

6

7

8

ORANGE
Seville Orange, *Citrus aurantium* (L.)
Watercolours and bodycolours within a ruled red ink border, and extending beyond on the right (3 mm).
21.5 x 14.5 cm (8 3/8 x 5 5/8 in.) 1962-7-14-1(44)

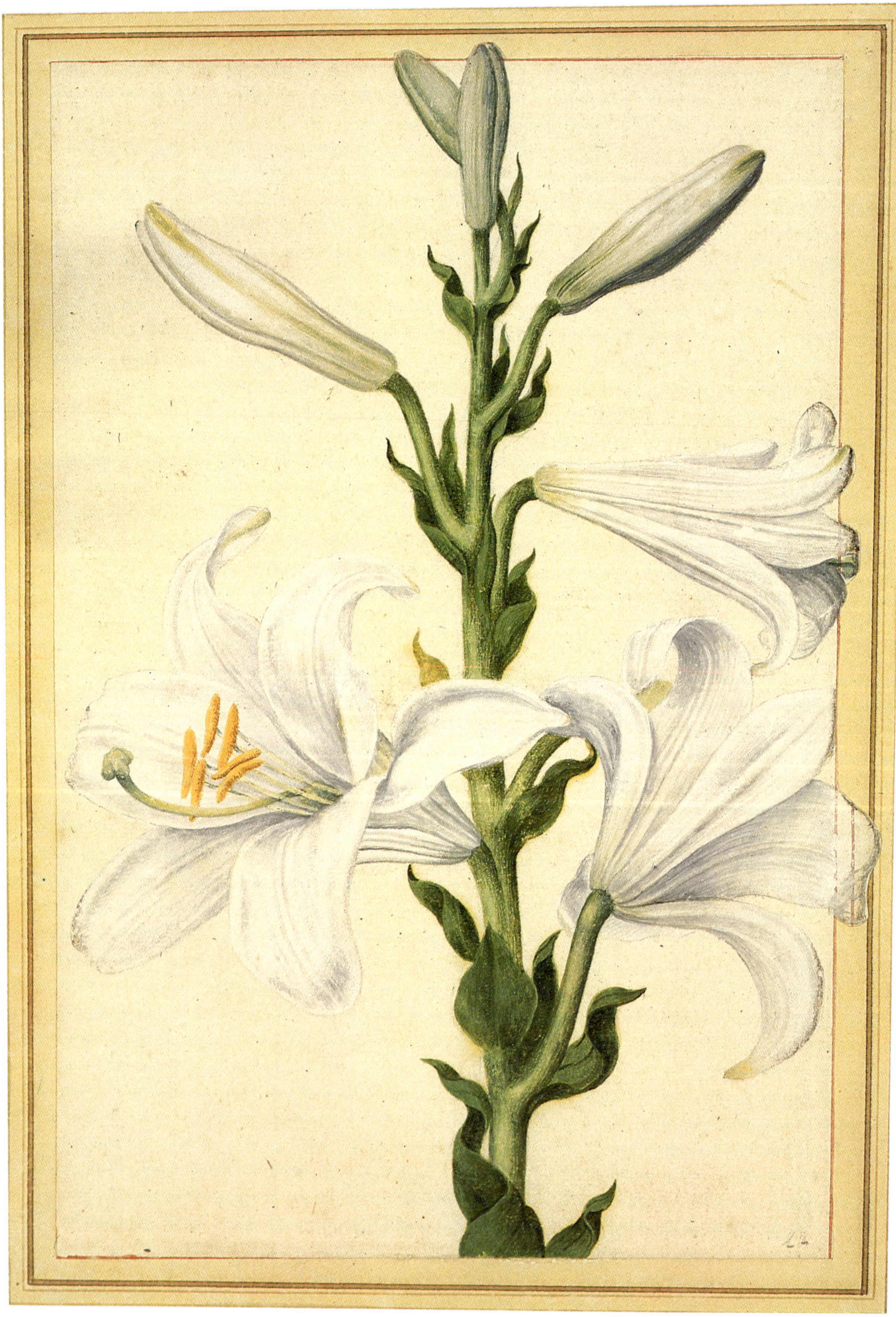

April

9

10

11

12

13

14

15

MADONNA LILY

Madonna Lily, *Lilium candidum* (L.)

Watercolours and bodycolours with some oxidization nearest the edges; within a ruled red ink border, largely cut away at the left and bottom but with a bud extending beyond it (3 mm) at the top and flower to the right (3 to 8 mm).

21.3 x 13.6 cm (8 3/8 x 5 3/8 in.) 1962-7-14-1(24)

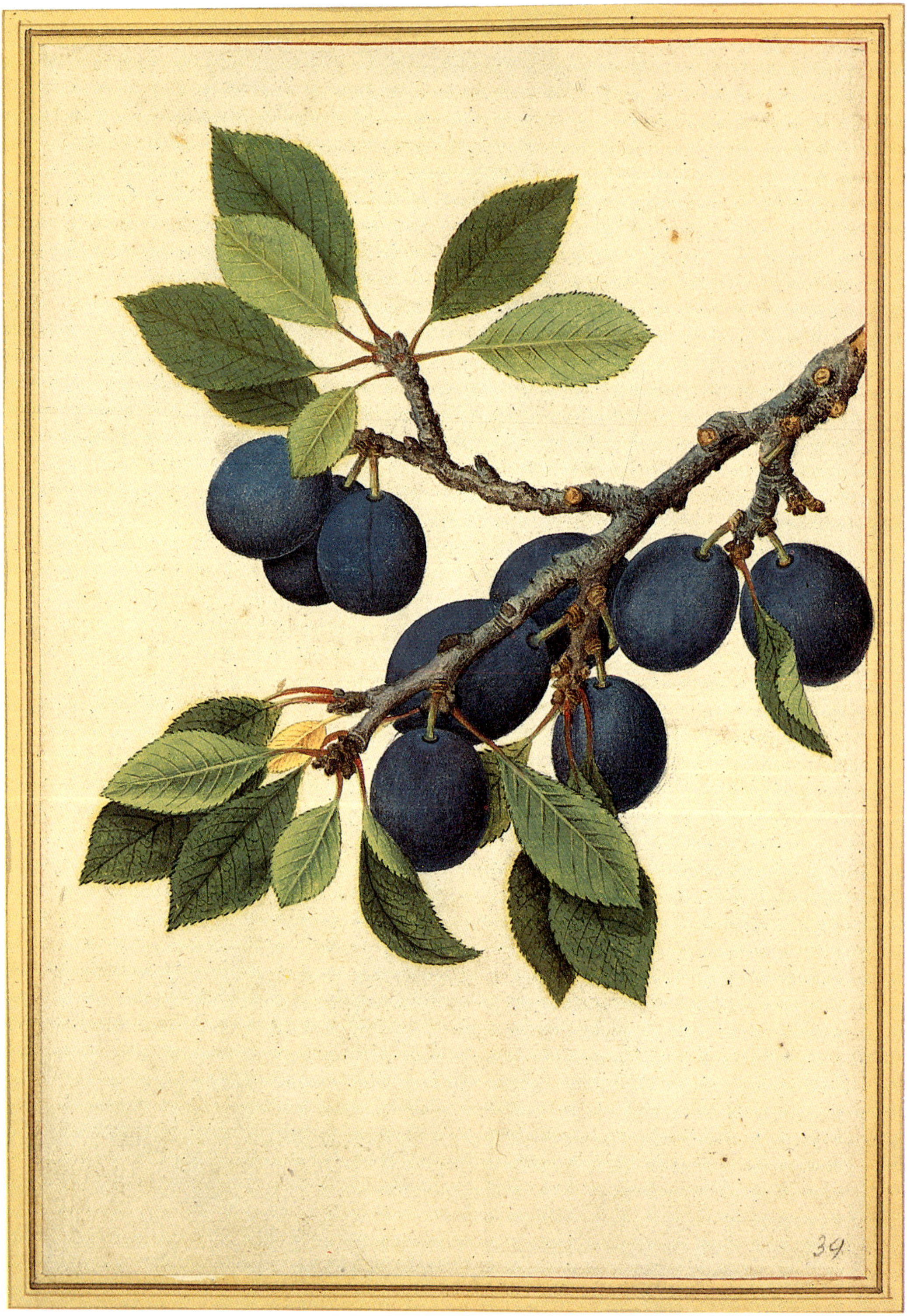
39

April

16

17

18

19

20

21

22

BULLACE

Wild Plum (Bullace), *Prunus domestica* (L.), subsp. *insititia* (L.) C.K. Schneider

Watercolours and bodycolours within a ruled red ink border, largely cut away at the sides and bottom and extending beyond it on the right (2 mm).

21.8 x 14.6 cm (8 1/2 x 5 3/4 in.) 1962-7-14-1(41)

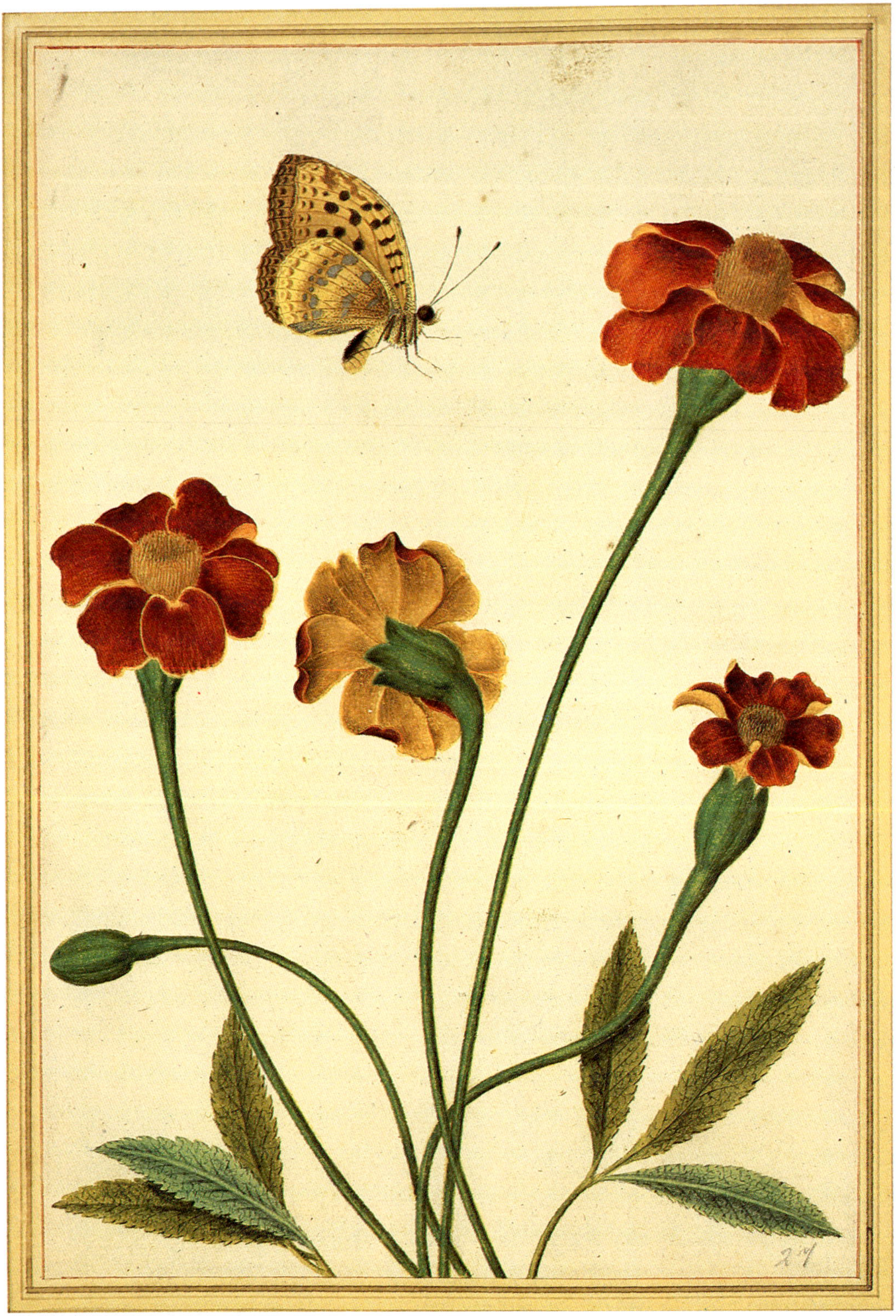

April

23

24

25

26

27

28

29

French Marigold and Butterfly
French Marigold, *Tagetes patula* (L.);
High Brown Fritillary Butterfly, *Argynnis adippe* (Denis and Schiffermüller)
Watercolours and bodycolours within a ruled red ink border.
21.3 x 14 cm (8 3/8 x 5 1/2 in.) 1962-7-14-1(29)

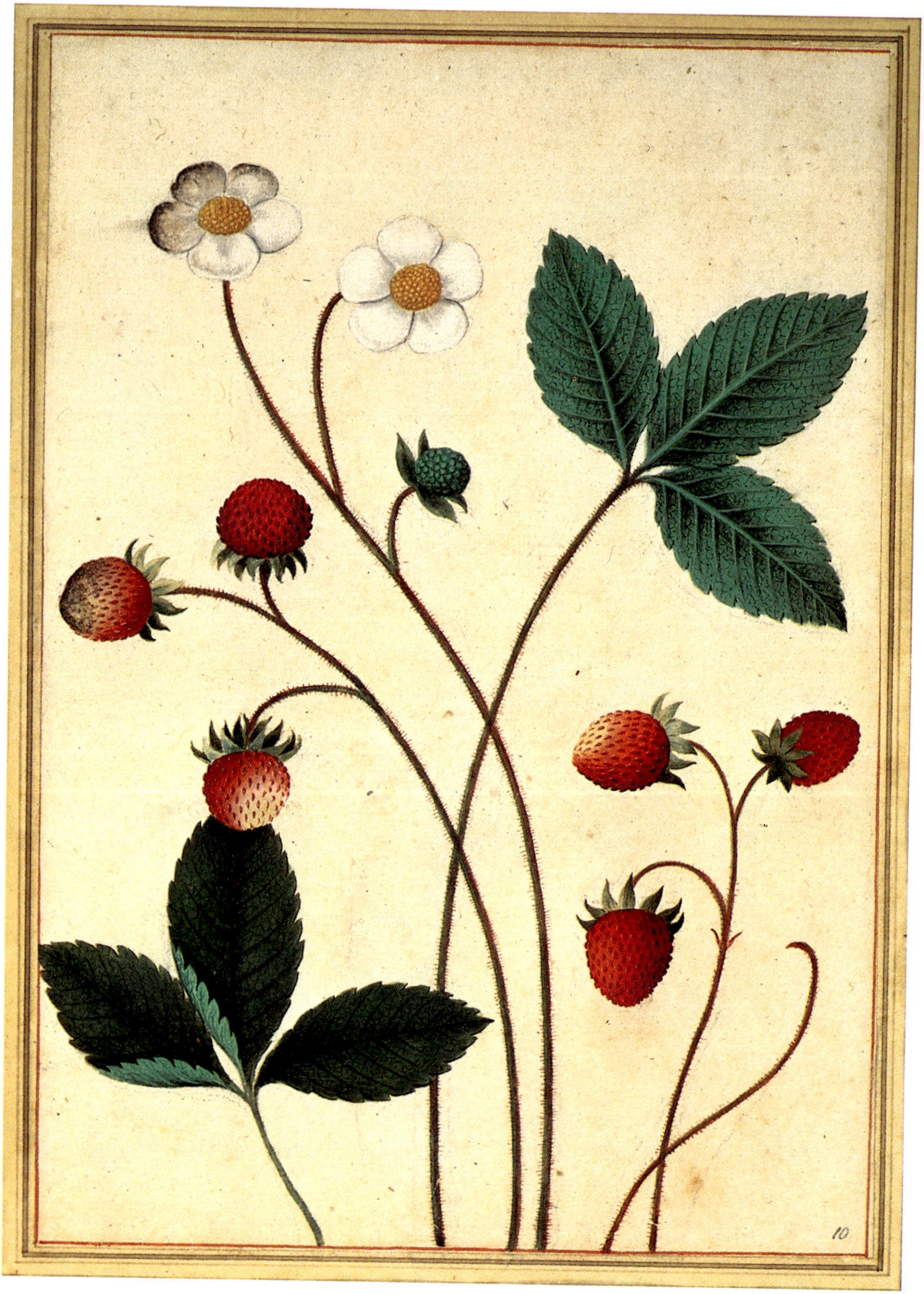
10

April/May

30

1

2

3

4

5

6

STRAWBERRY
Hautbois Strawberry, *Fragaria moschata* Duchesne
Watercolours and bodycolours with slight oxidization on the top flower, within a ruled red ink border, partly cut away on the right.
21.2 x 14.7 cm (8 3/8 x 5 3/4 in.) 1962-7-14-1(12)

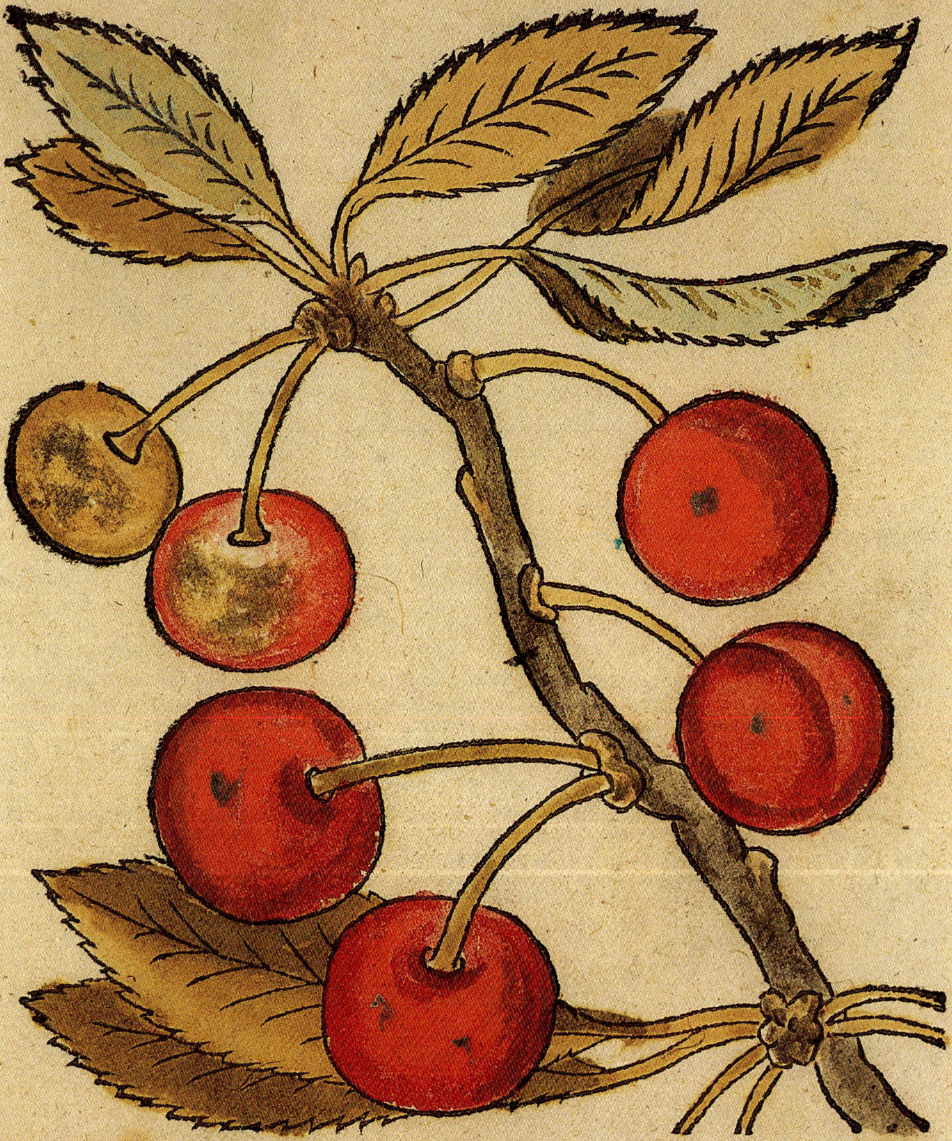

Latinè CERASA.
Gallicé CERISES.
Anglicé CHERIES.

May

7

8

9

10

11

12

13

WILD CHERRY (detail)
La Clef des Champs, 1586
Woodcut, hand-coloured
7.6 x 6.2 cm (3 x 2 3/8 in.)
Sloane Min. 269

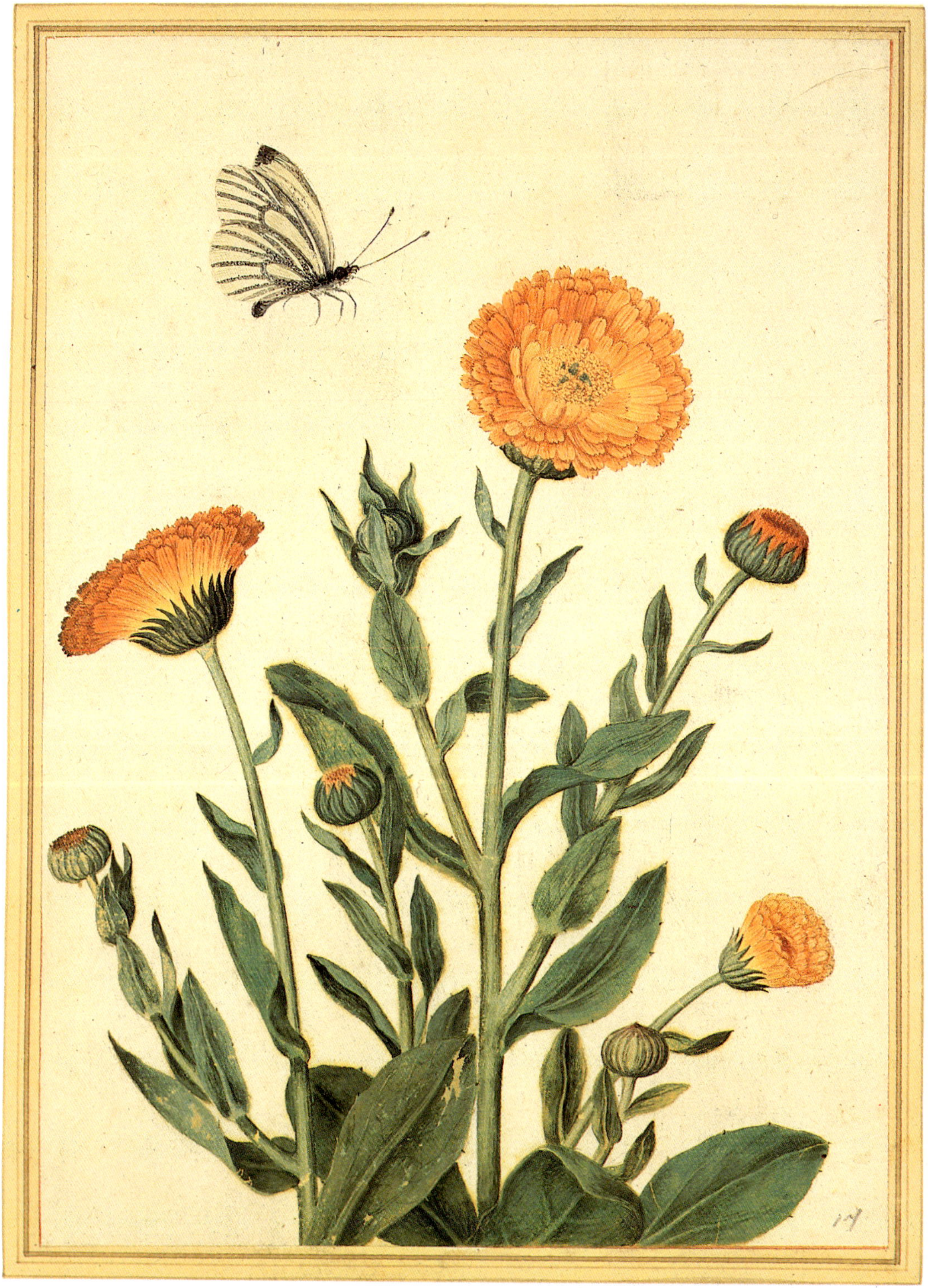

May

14

15

16

17

18

19

20

Pot Marigold and Green-veined White Butterfly
Pot Marigold, *Calendula officinalis* (L.); Green-veined White Butterfly, *Pieris napi* (L.)
Watercolours and bodycolours with slight flaking on the leaves; within a ruled red ink border, largely cut away on all sides.
20.7 x 14.5 cm (8 1/8 x 5 3/4 in.) 1962-7-14-1(19)

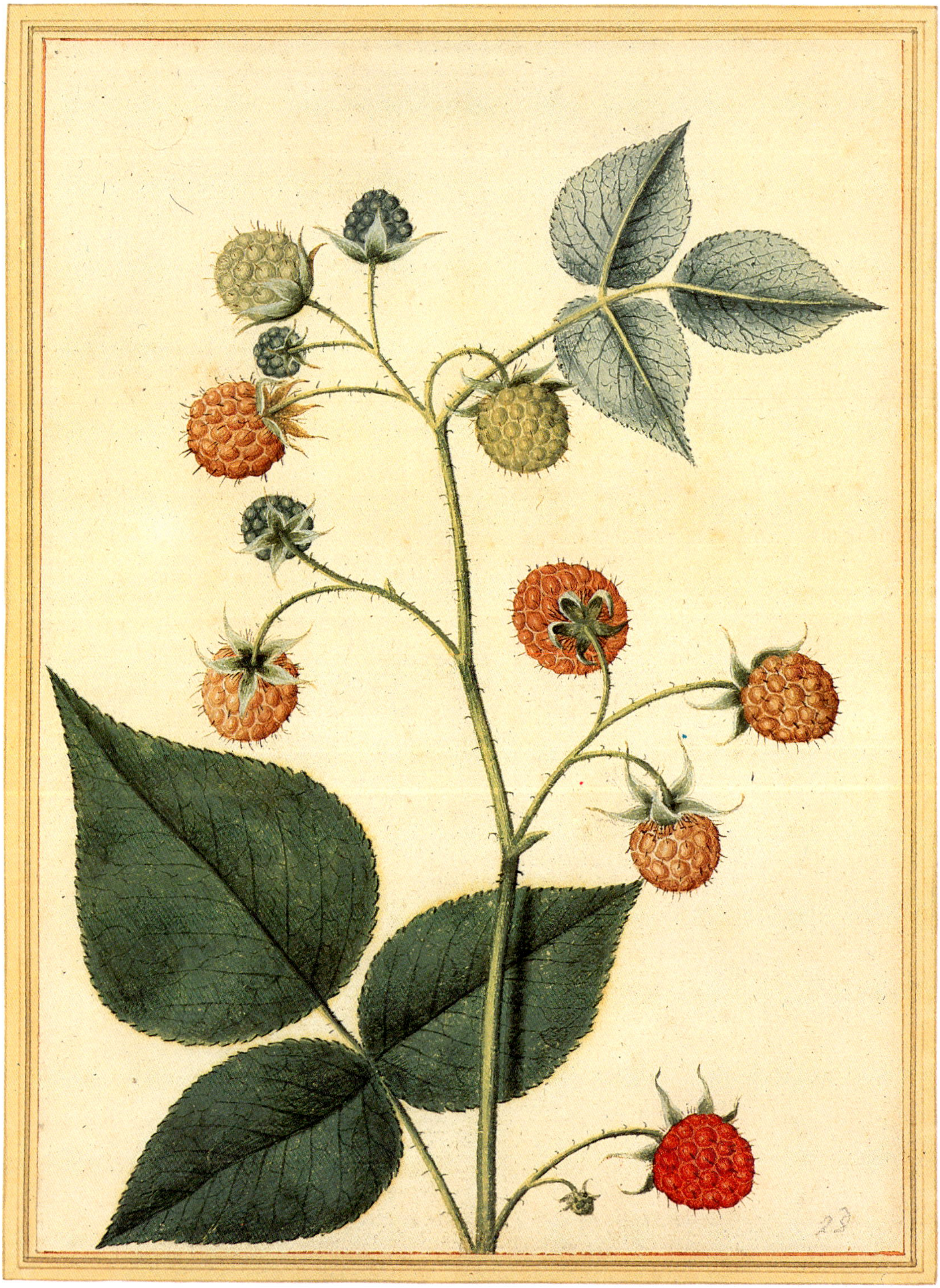

May

21

22

23

24

25

26

27

RASPBERRY
Raspberry, *Rubus idaeus* (L.)
Watercolours and bodycolours within a ruled red ink border, partly cut away at the top and left.
20.9 x 14.8 cm (8 1/4 x 5 3/4 in.) 1962-7-14-1(30)

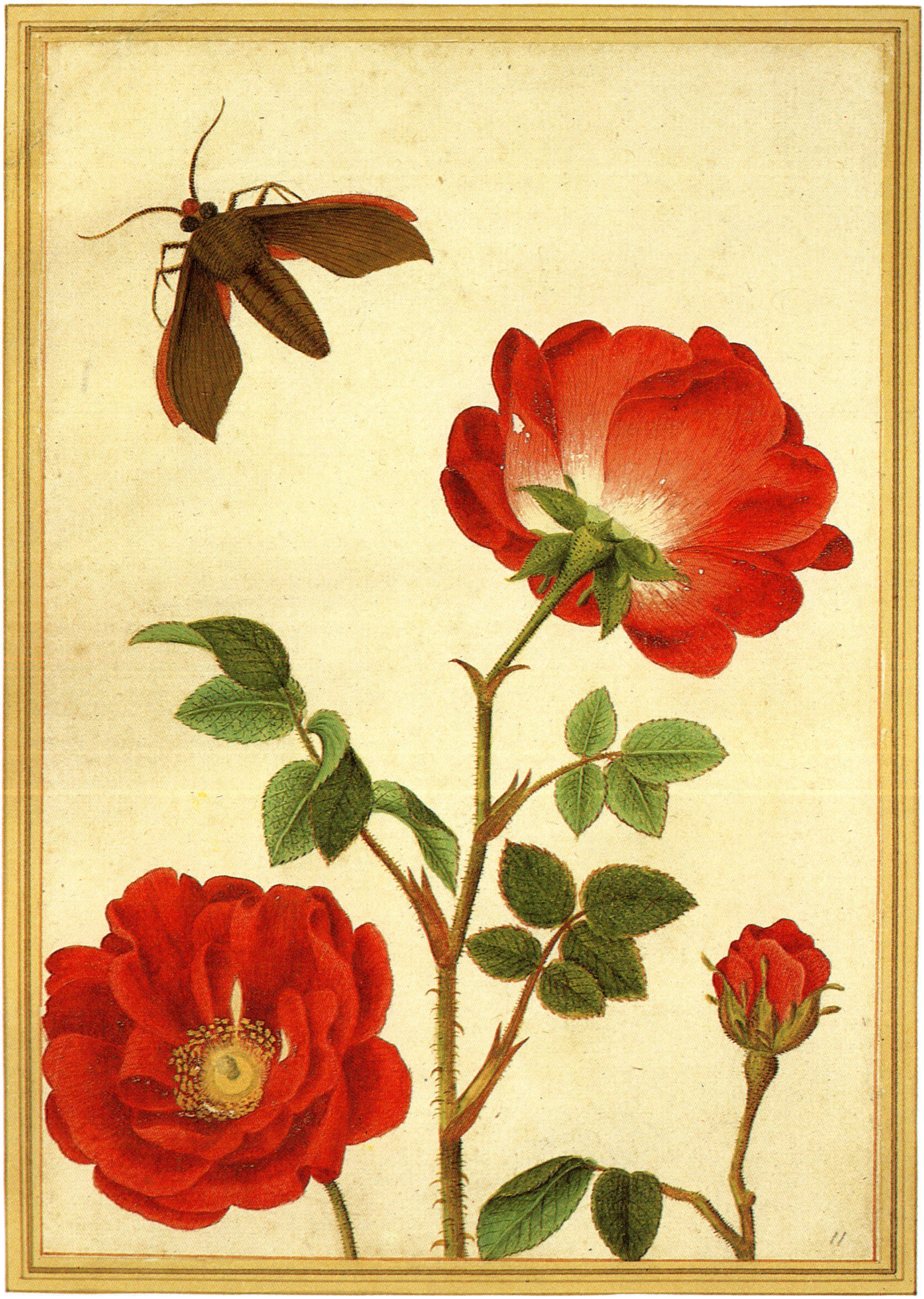

May/June

28

29

30

31

1

2

3

FRENCH ROSE AND PRIVET HAWK MOTH
French Rose, *Rosa gallica* (L.); Privet Hawk Moth, *Sphinx ligustri* (L.)
Watercolours and bodycolours with slight flaking of colour on the upper flower, within a ruled red ink border, partly cut away at the top and right.
21.1 x 14.5 cm (8 1/4 x 5 3/4 in.) 1962-7-14-1(13)

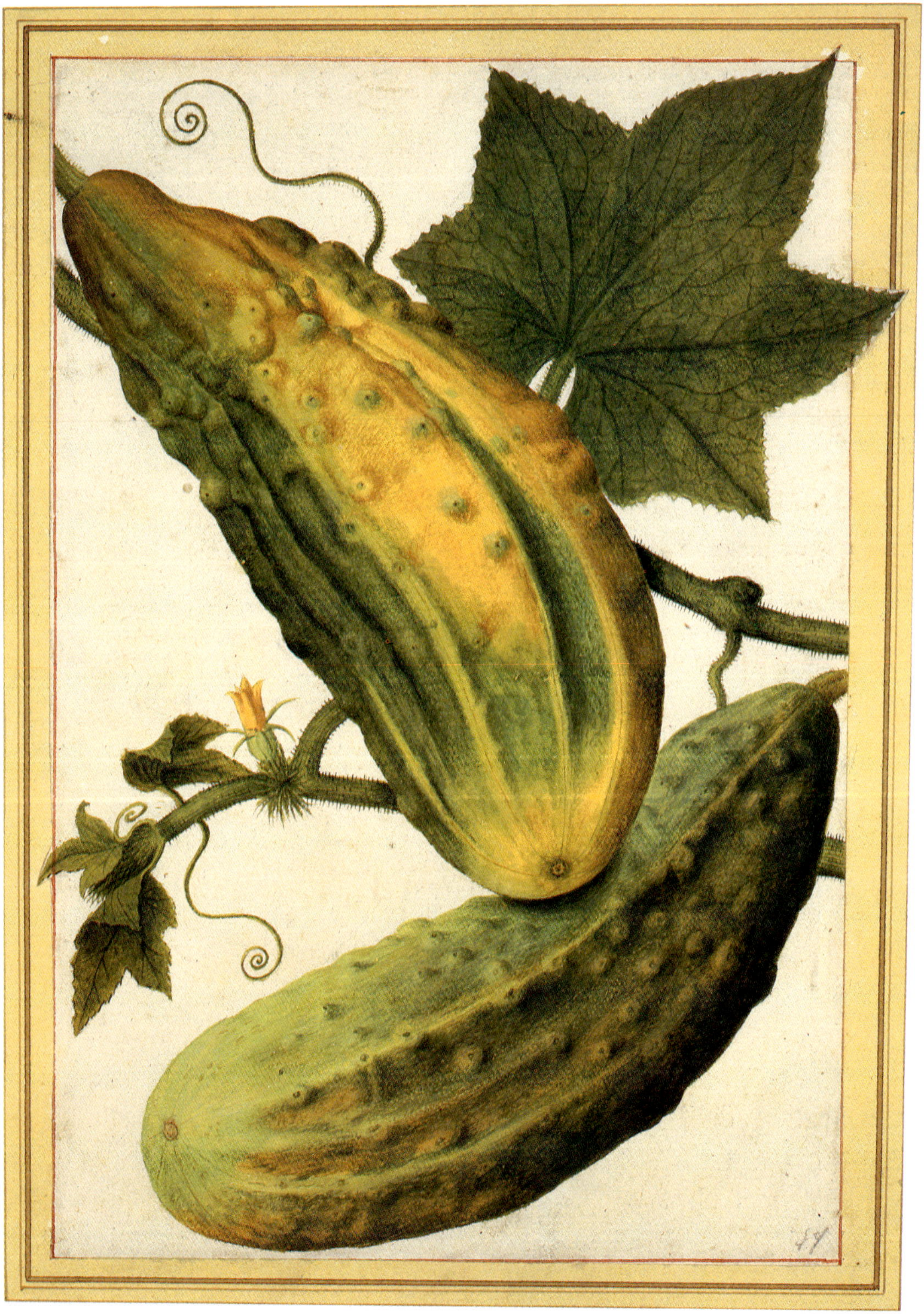

June

4

5

6

7

8

9

10

CUCUMBER
Cucumber, *Cucumis sativus* (L.)
Watercolours and bodycolours within a ruled red ink border, and extending beyond it by 3 mm at the top and the left and 10 mm at the right.
21.5 x 14 cm (8 1/2 x 5 1/2 in.) 1962-7-14-1(49)

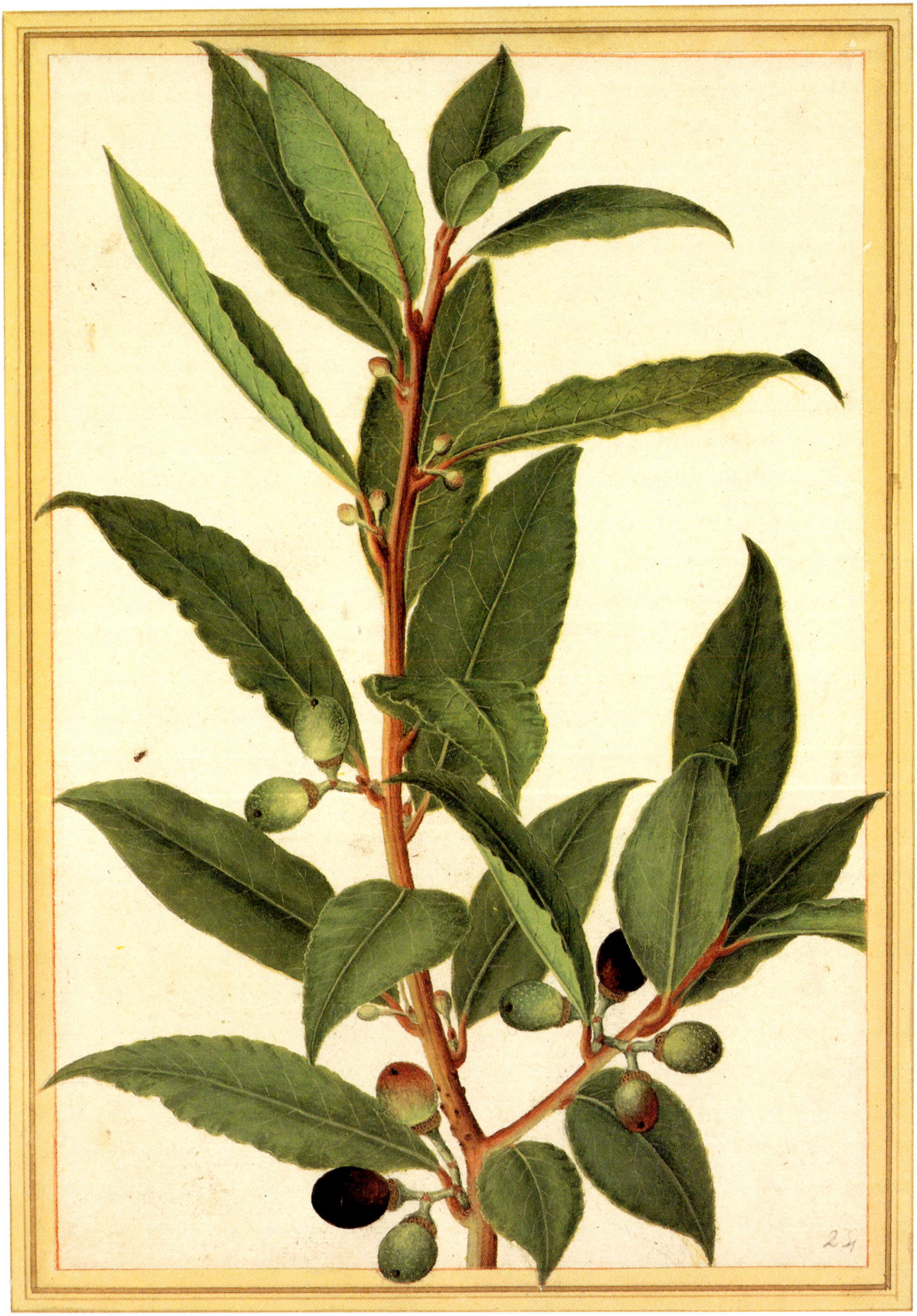

June

11

12

13

14

15

16

17

SWEET BAY
Laurel (Sweet Bay), *Laurus nobilis* (L.)
Watercolours and bodycolours within a ruled red ink border, and extending beyond it at each edge (up to 4 mm).
21.2 x 14.1 cm (8⅜ x 5½ in.) 1962-7-14-1(26)

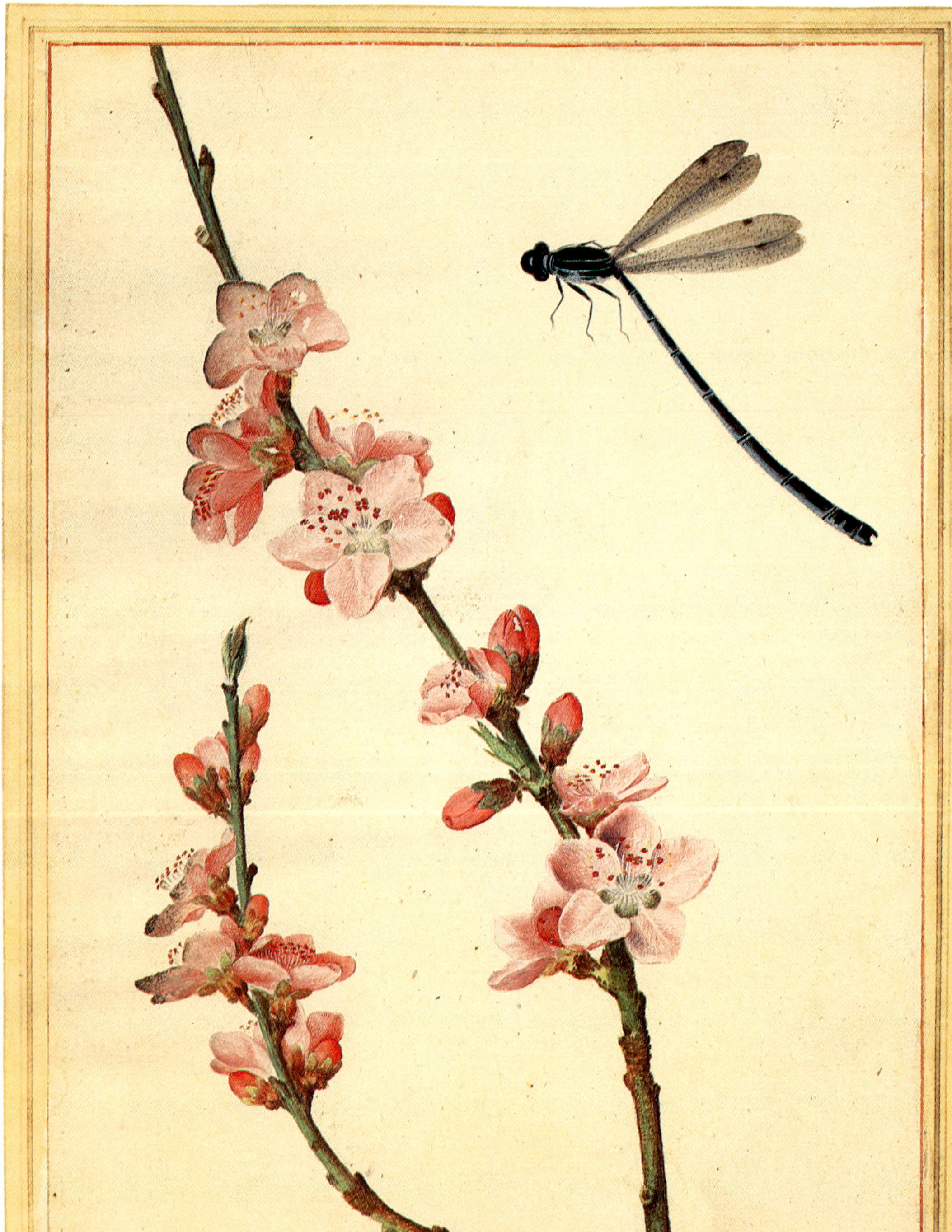

June

18

19

20

21

22

23

24

PEACH BLOSSOM AND DAMSELFLY
Peach, *Prunus persica* (L.) Batsch; Damselfly, female of the *Coenegrion* sp.
Watercolours and bodycolours with some flaking on the petals and slight oxidization of the petals nearest the borders, within a ruled red ink border, partly cut away at the bottom.
20.5 x 14.7 cm (8 1/4 x 5 3/4 in.) 1962-7-14-1(6)

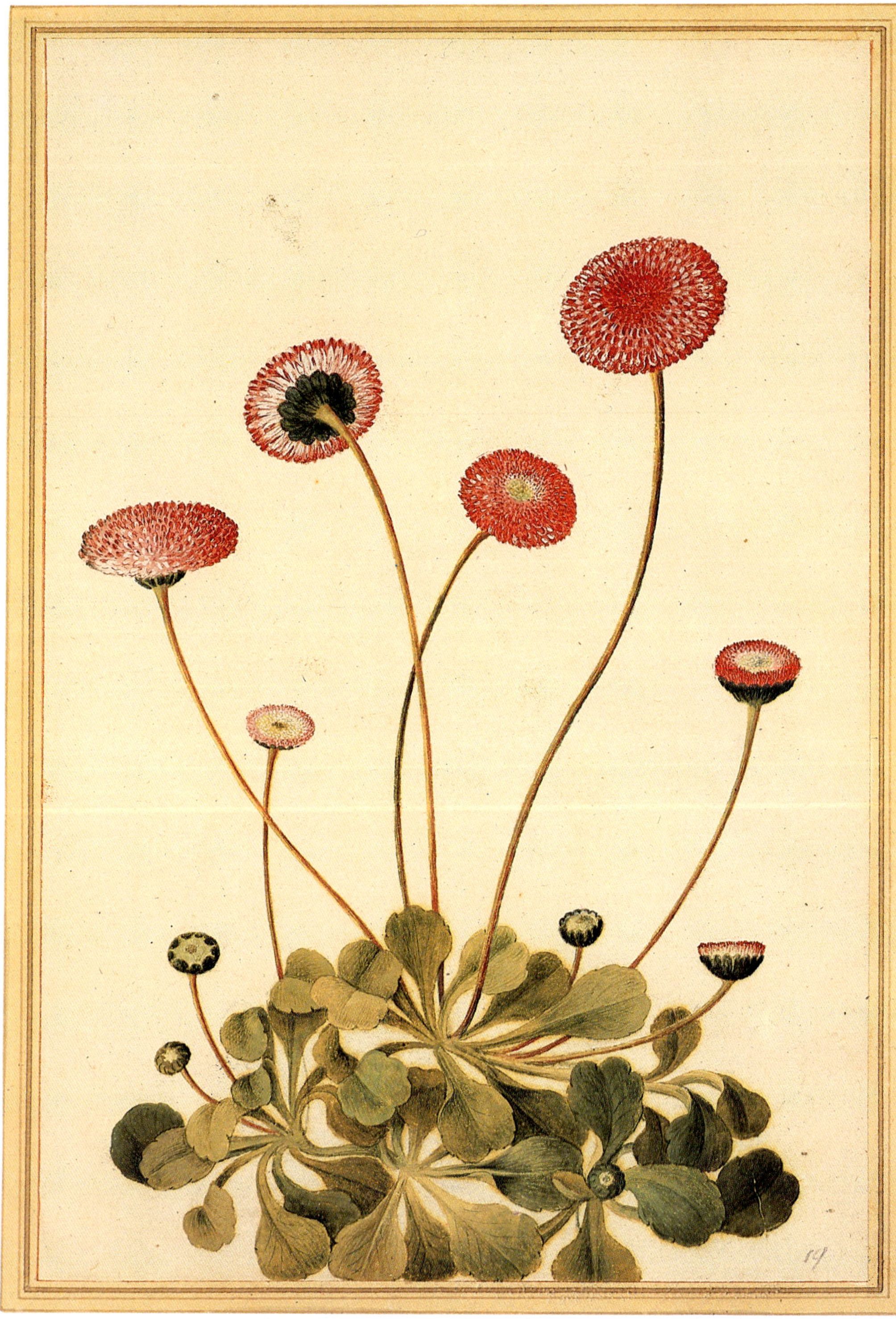
14

June/July

25

26

27

28

29

30

1

DAISY
Double Daisy, *Bellis perennis* (L.)
Watercolours and bodycolours within a ruled red ink border, partly cut away at the top.
21.4 x 14.3 cm ($8\frac{3}{8}$ x $5\frac{5}{8}$ in.) 1962-7-14-1(21)

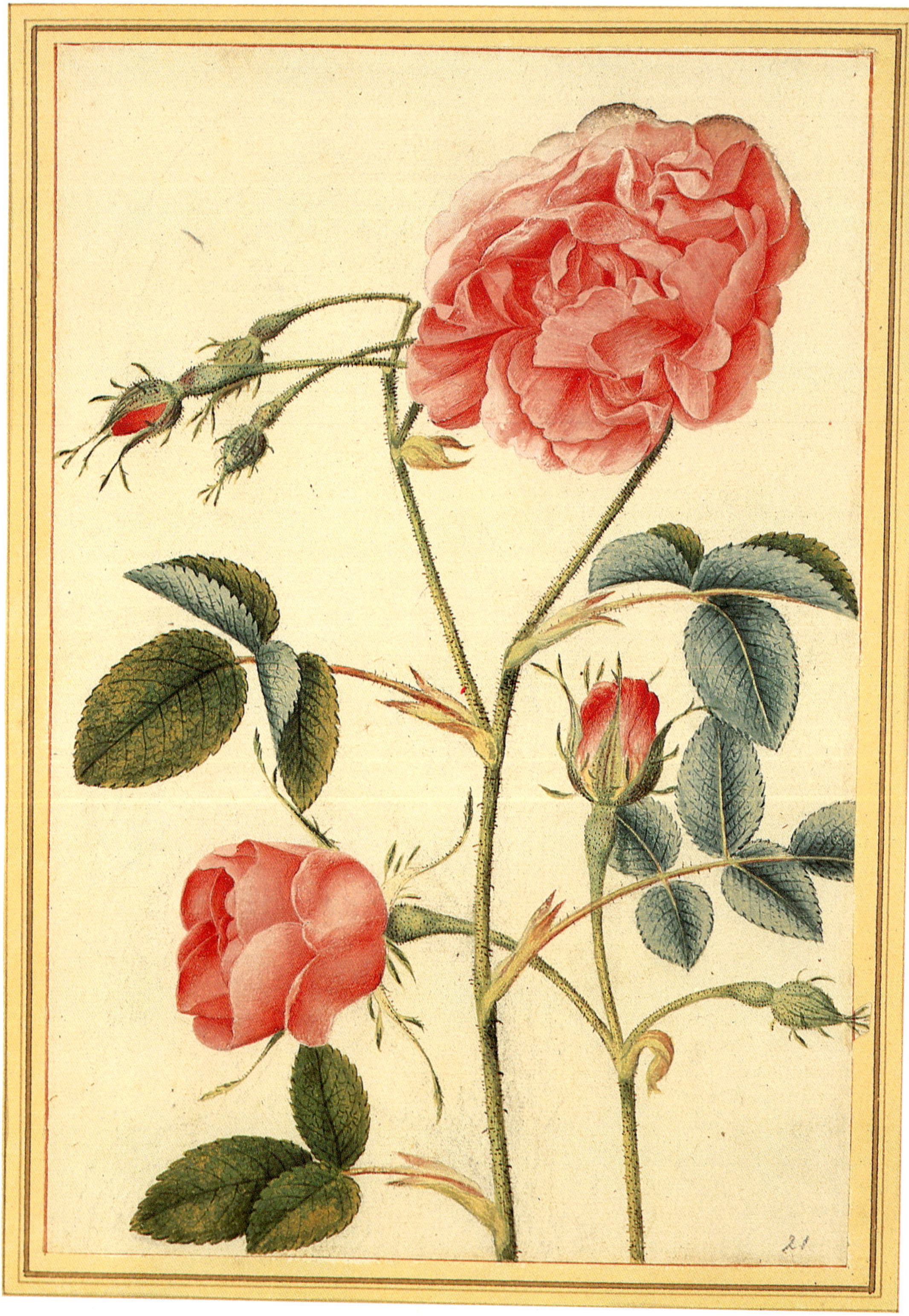
21

July

2

3

4

5

6

7

8

CABBAGE ROSE
Cabbage Rose, *Rosa centifolia* (L.)
Watercolours and bodycolours with slight oxidisation near the top border; within a ruled red ink border, partly cut away at the right and bottom.
21.3 x 14 cm (8 3/8 x 5 1/2 in.) 1962-7-14-1(23)

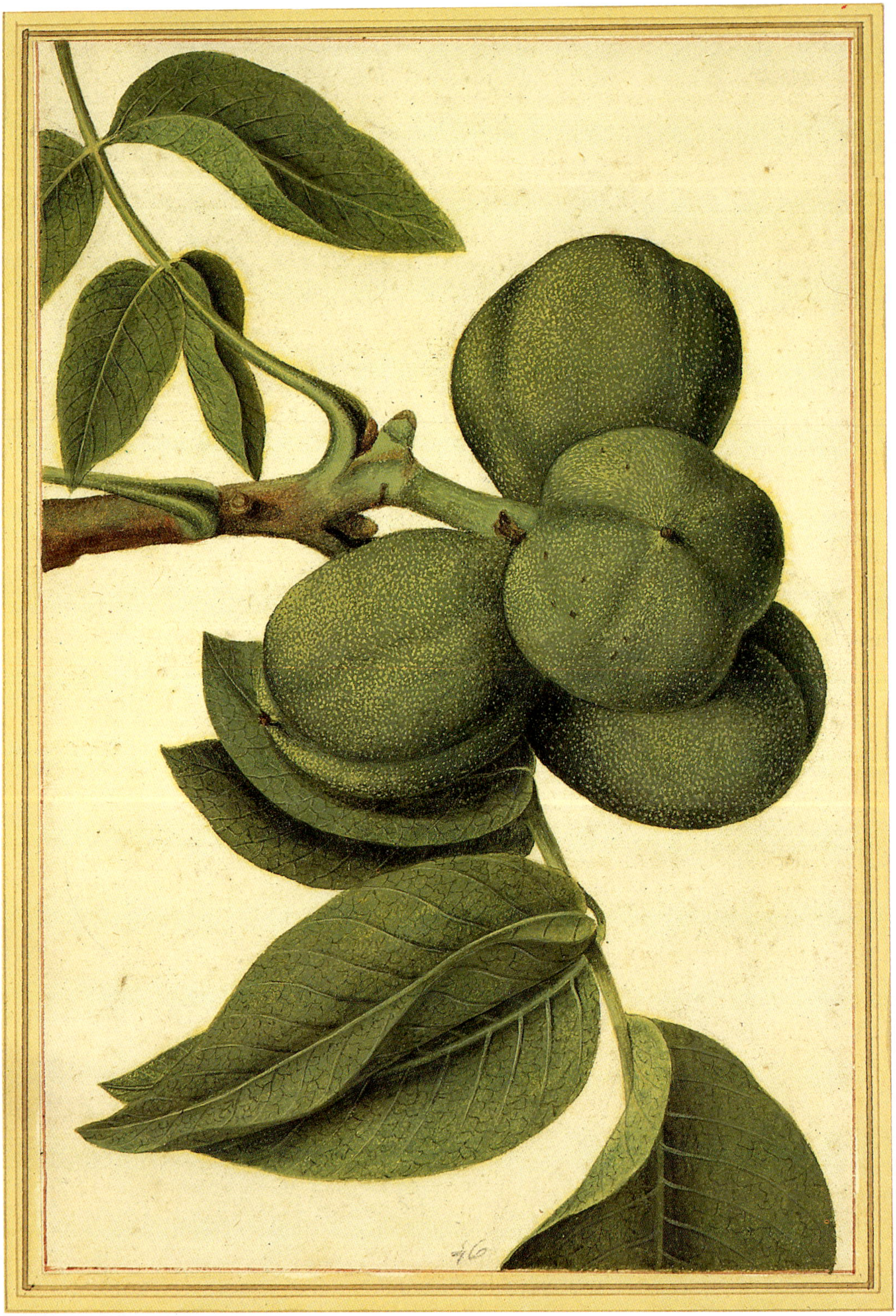
46

July

9

10

11

12

13

14

15

WALNUT
Walnut, *Juglans regia* (L.)
Watercolours and bodycolours within a ruled red ink border.
21.5 x 14 cm (8½ x 5½ in.) 1962-7-14-1(48)

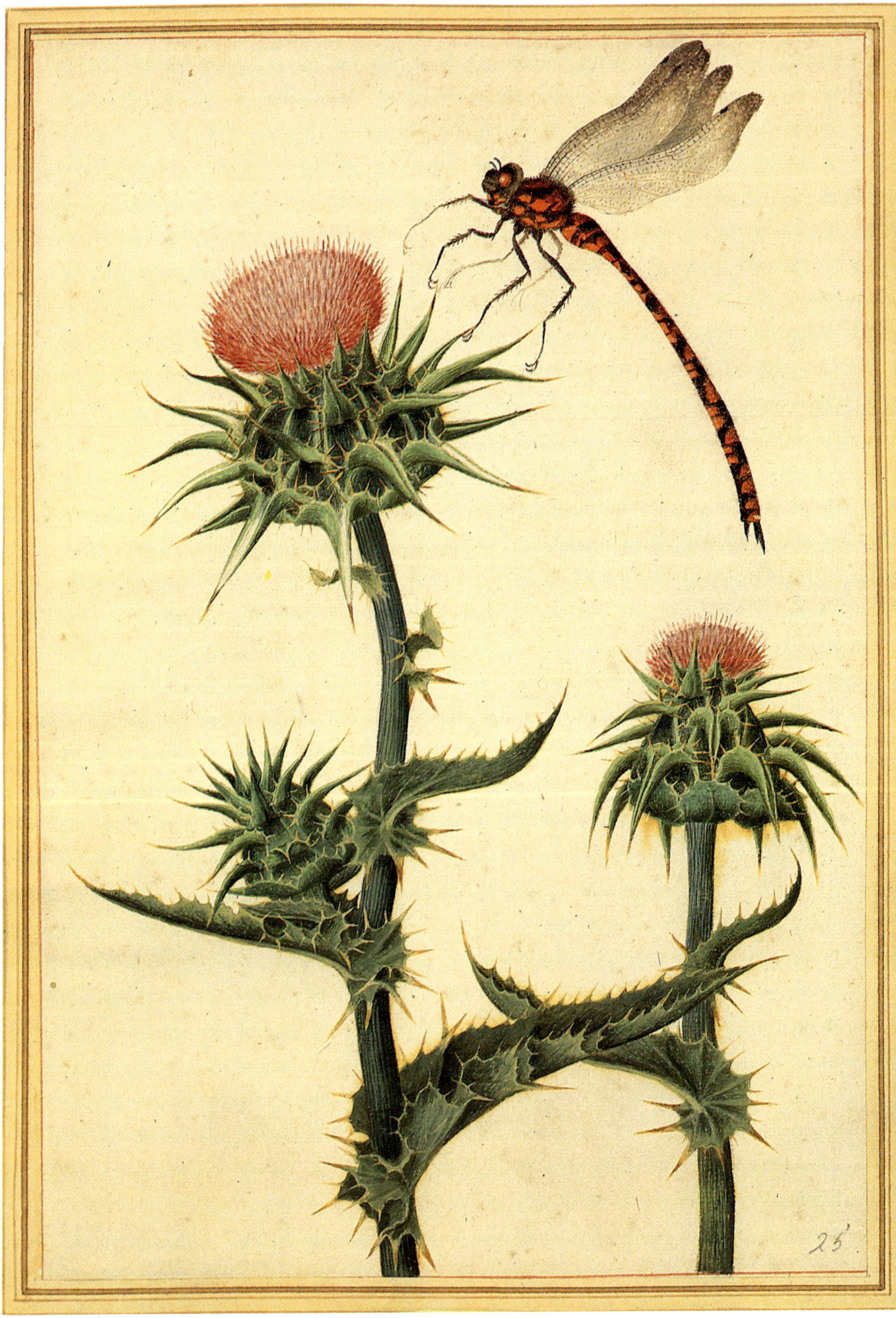
25

July

16

17

18

19

20

21

22

THISTLE AND DRAGONFLY
Milk Thistle or Saint Mary's Thistle, *Silybum marianum* (L.) Gaertner;
Dragonfly, possibly *Aeshna* sp.
Watercolours and bodycolours with some oxidization on the insect's wings;
within a ruled red ink border, partly cut away at the right and bottom.
21.4 x 14.1 cm (8 3/8 x 5 1/2 in.) 1962-7-14-1(27)

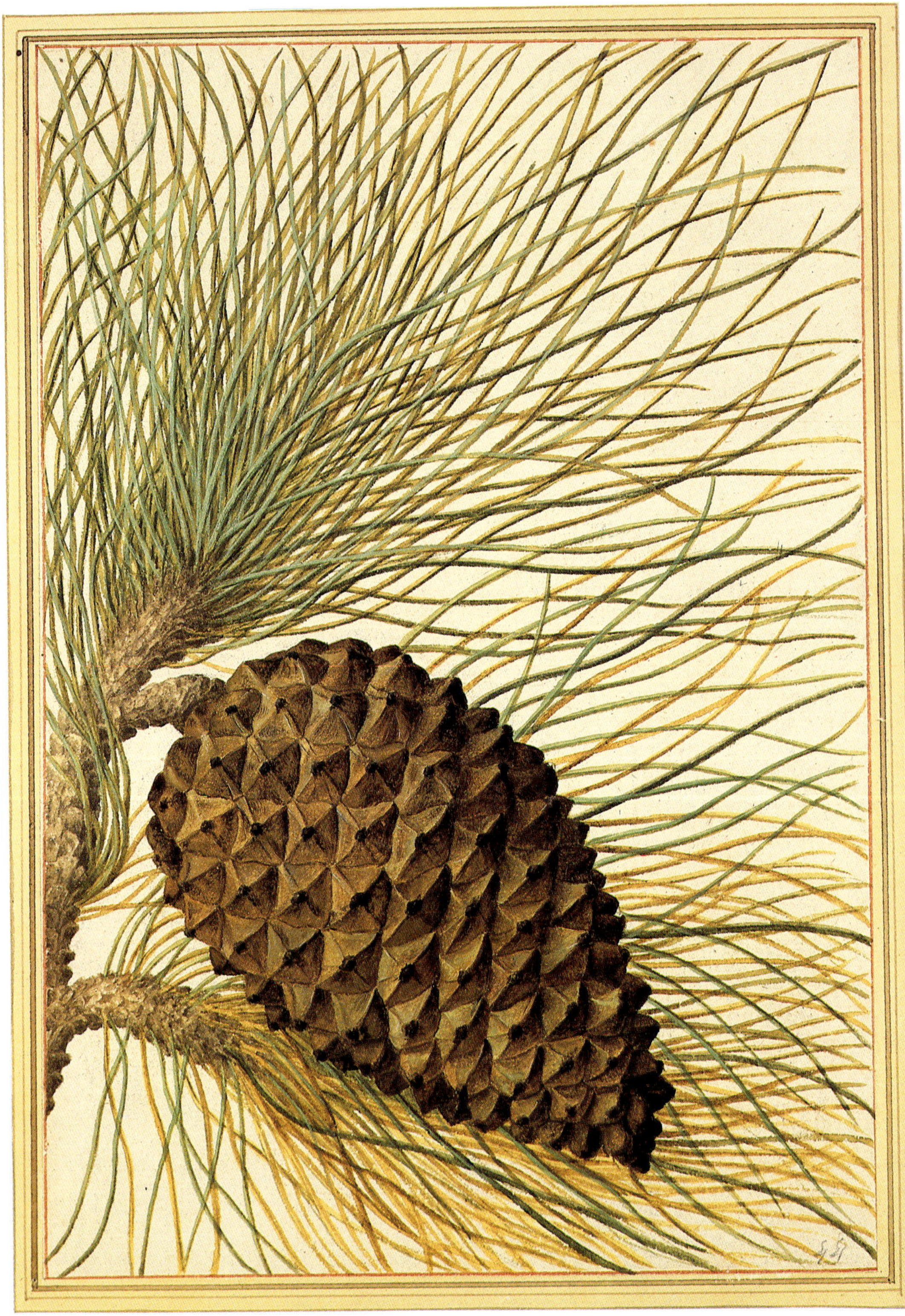

July

23

24

25

26

27

28

29

PINE
Maritime Pine, *Pinus pinaster* Aiton
Watercolours and bodycolours within a ruled red ink border.
21.5 x 14.5 cm (8 1/2 x 5 5/8 in.) 1962-7-14-1(46)

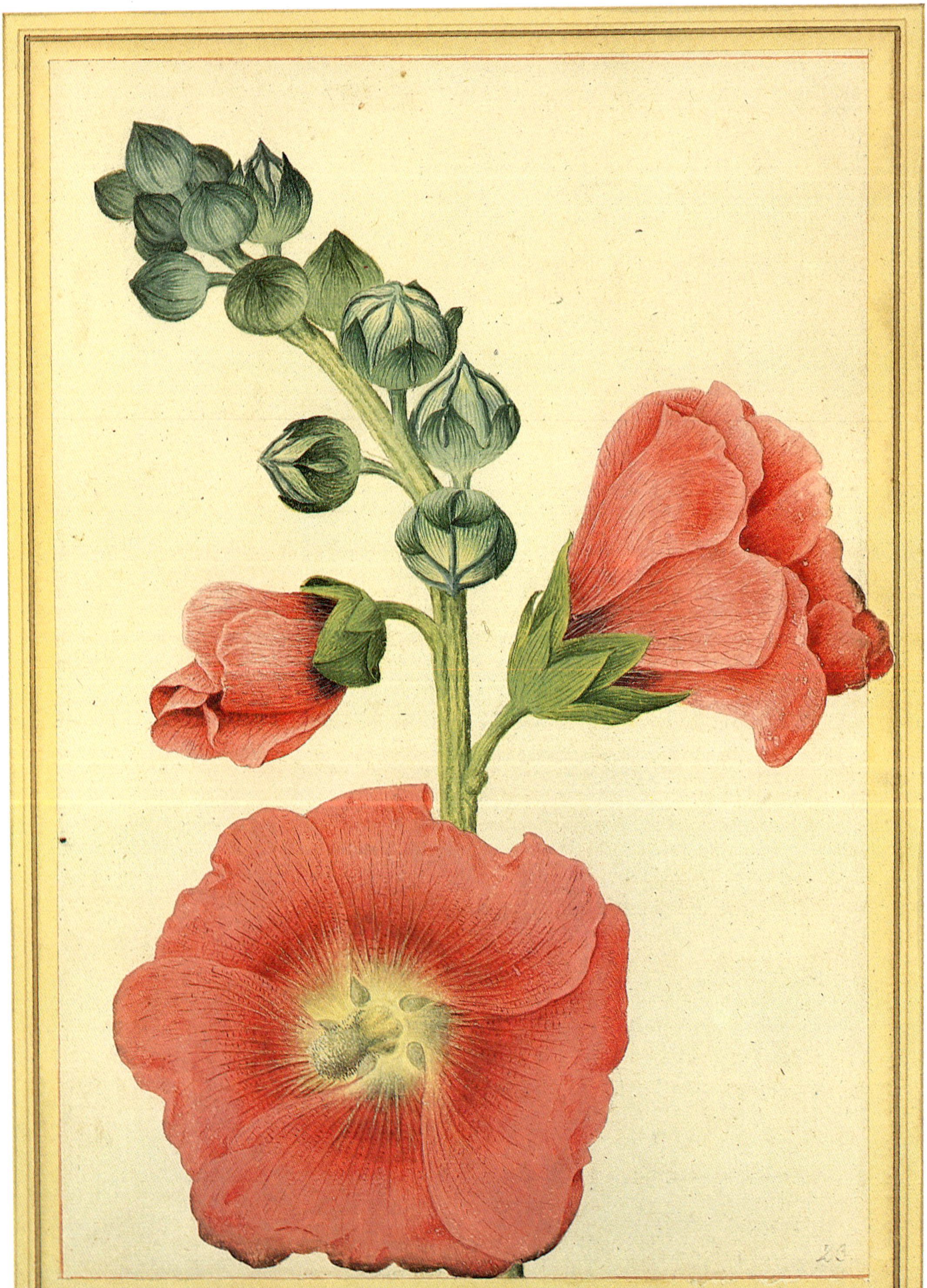

July/August

30

31

1

2

3

4

5

HOLLYHOCK
Hollyhock, *Althaea rosea* (L.) Cavanilles
Watercolours and bodycolours with slight flaking on the flowers; within a ruled red ink border, partly cut away on the left and on the right and extending beyond it on the right (5mm).
21.4 x 14 cm ($8\frac{3}{8}$ x $5\frac{1}{2}$ in.) 1962-7-14-1(25)

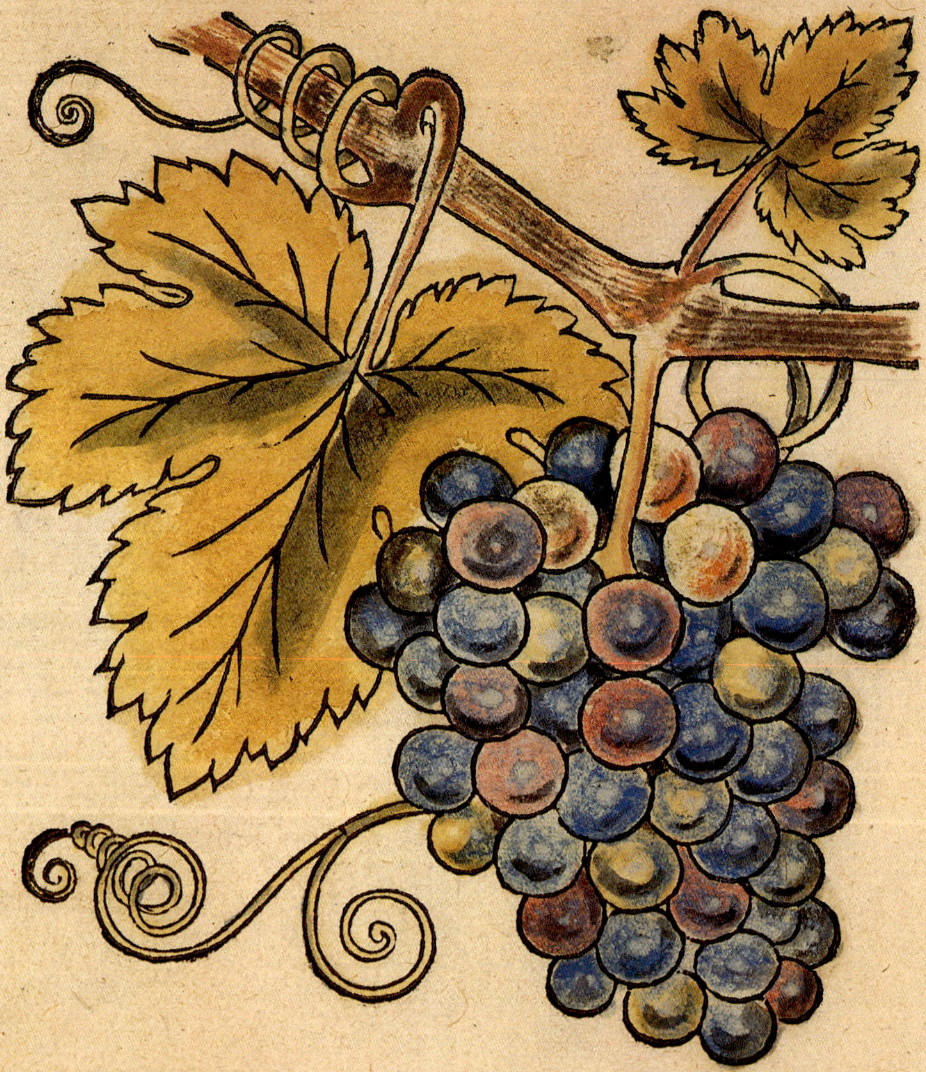

Latinè	VITIS VINIFERA.
Gallicé	VIGNE.
Anglicé	GRAPES.

August

6

7

8

9

10

11

12

GRAPEVINE (detail)
La Clef des Champs, 1586
Woodcut, hand-coloured
8.3 x 7.2 cm (3 1/4 x 2 7/8 in.)
Sloane Min. 269

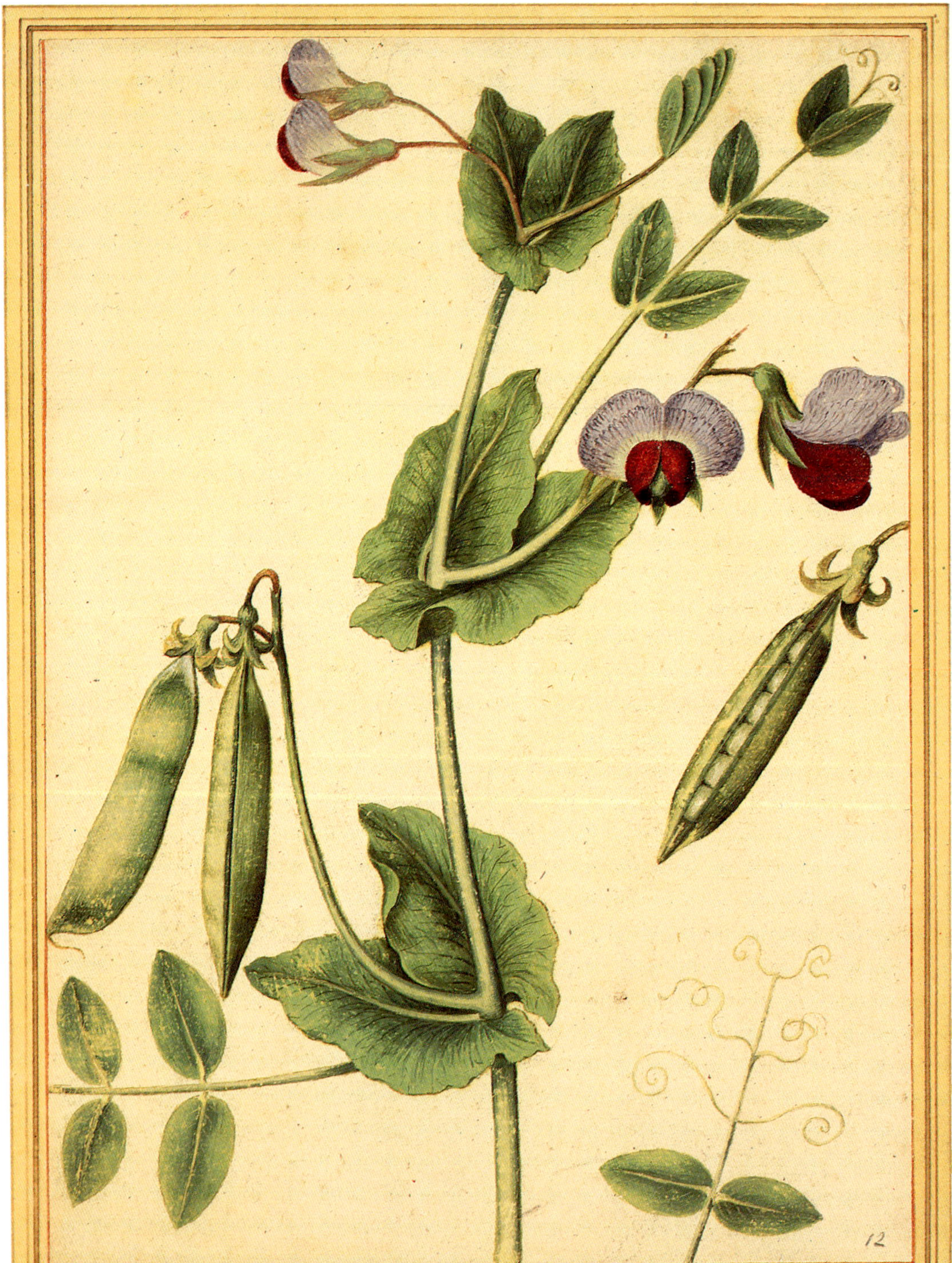
12

August

13

14

15

16

17

18

19

PEA
Field Pea, *Pisum sativum* (L.) var. *arvense* (L.) Poiret
Watercolours and bodycolours within a ruled red ink border, partly cut away at the top and bottom.
20.8 x 14.6 cm (8 1/4 x 5 3/4 in.) 1962-7-14-1(14)

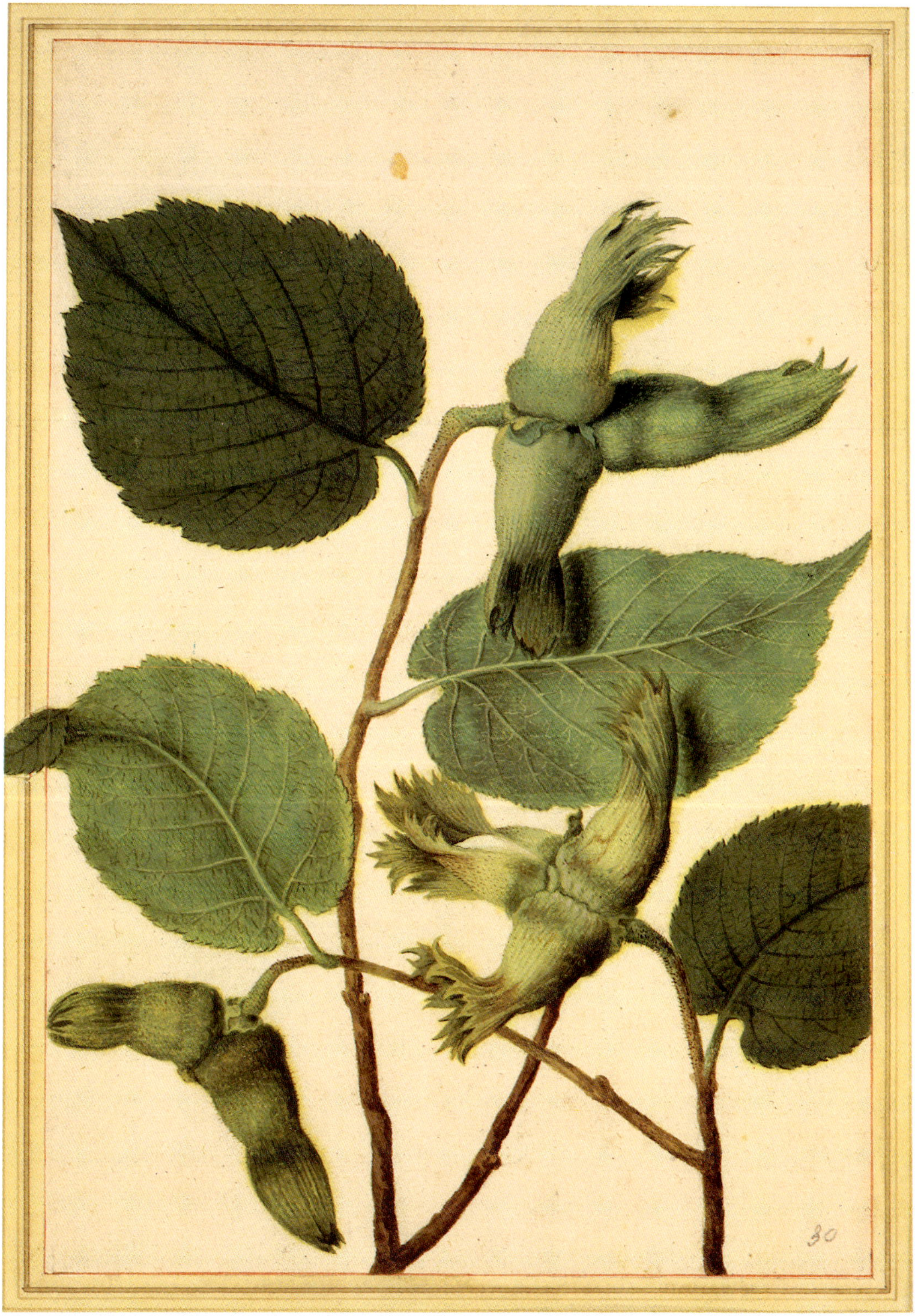
30

August

20

21

22

23

24

25

26

FILBERT
Filbert, *Corylus maxima* Miller
Watercolours and bodycolours with some yellowish staining from the edges of the nuts; within a ruled red ink border, but with a leaf extending beyond it on the left (18 mm).
21.4 x 14.3 cm (8 3/8 x 5 5/8 in.) 1962-7-14-1(32)

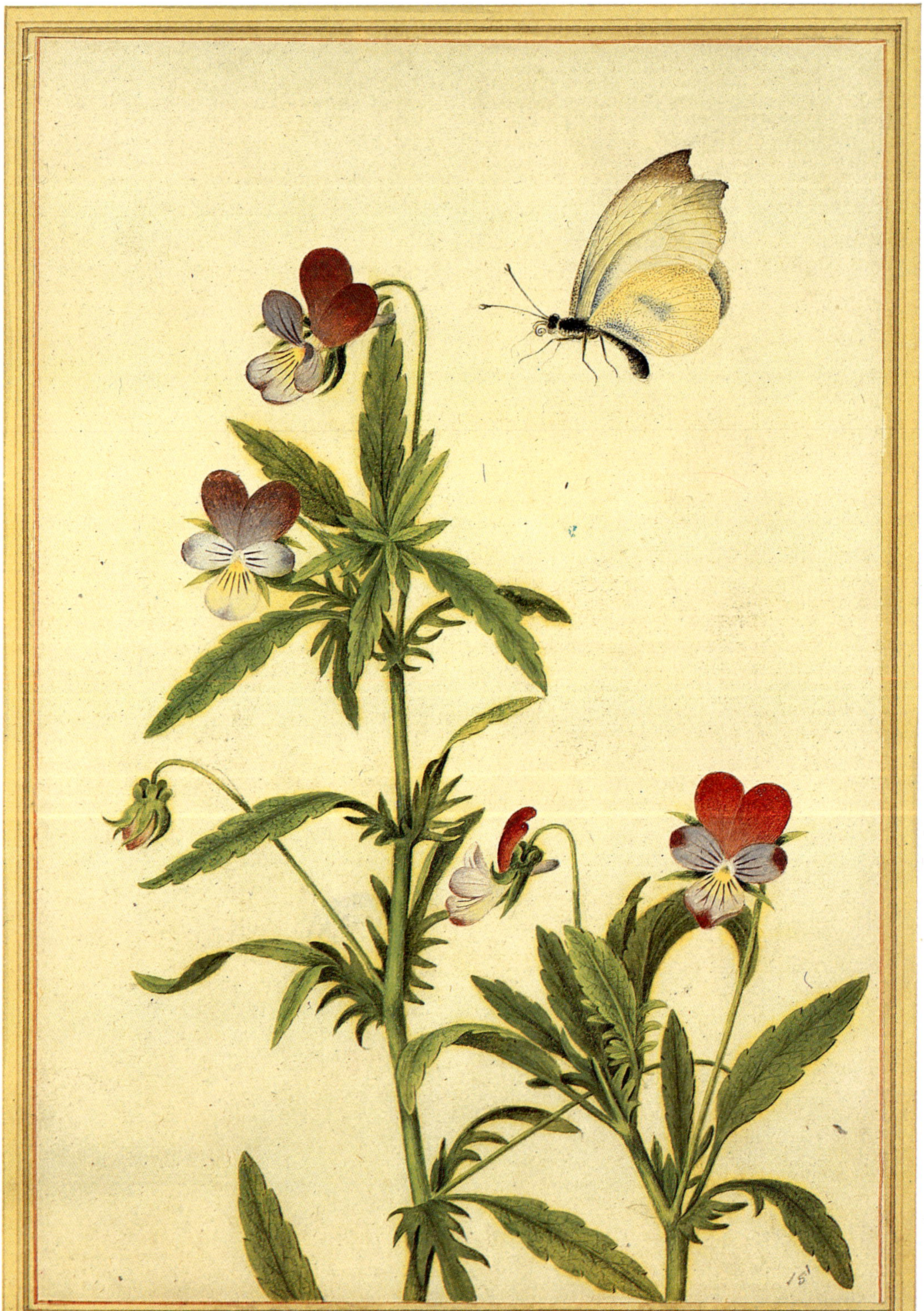

August/September

27

28

29

30

31

1

2

HEARTSEASE AND LARGE WHITE BUTTERFLY
Wild Pansy (Heartsease), *Viola tricolor* (L.); Large White Butterfly, *Pieris brassicae* (L.)
Watercolours and bodycolours with some oxidization on the butterfly's wings; within a ruled red ink border, partly cut away at the right and bottom.
21.4 x 14.3 cm (8 3/8 x 5 5/8 in.) 1962-7-14-1(17)

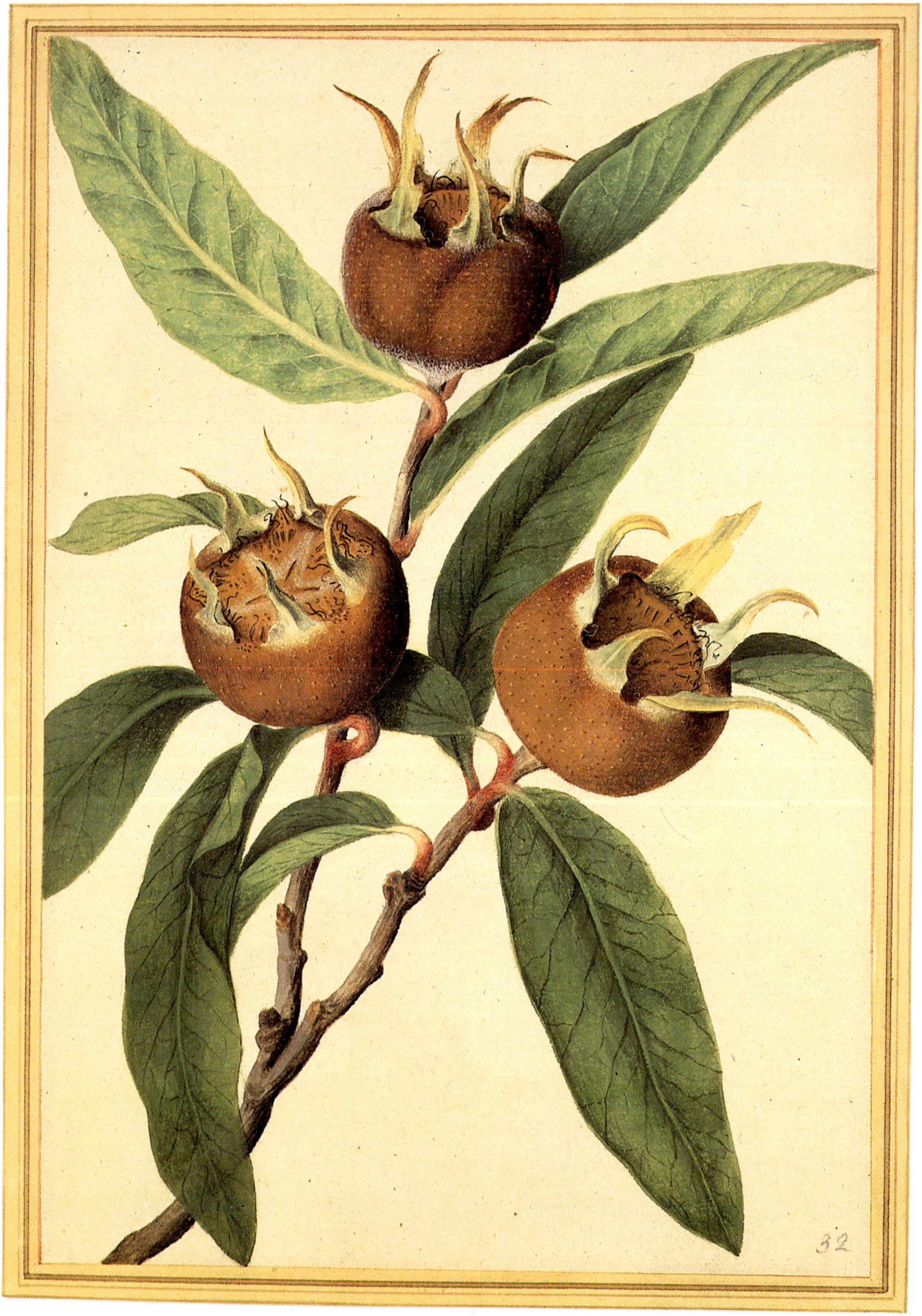
32

September

3

4

5

6

7

8

9

MEDLAR
Medlar, *Mespilus germanica* (L.)
Watercolours and bodycolours within a ruled red ink border, largely cut away at the bottom.
21.2 x 14.2 cm (8 3/8 x 5 5/8 in.) 1962-7-14-1(34)

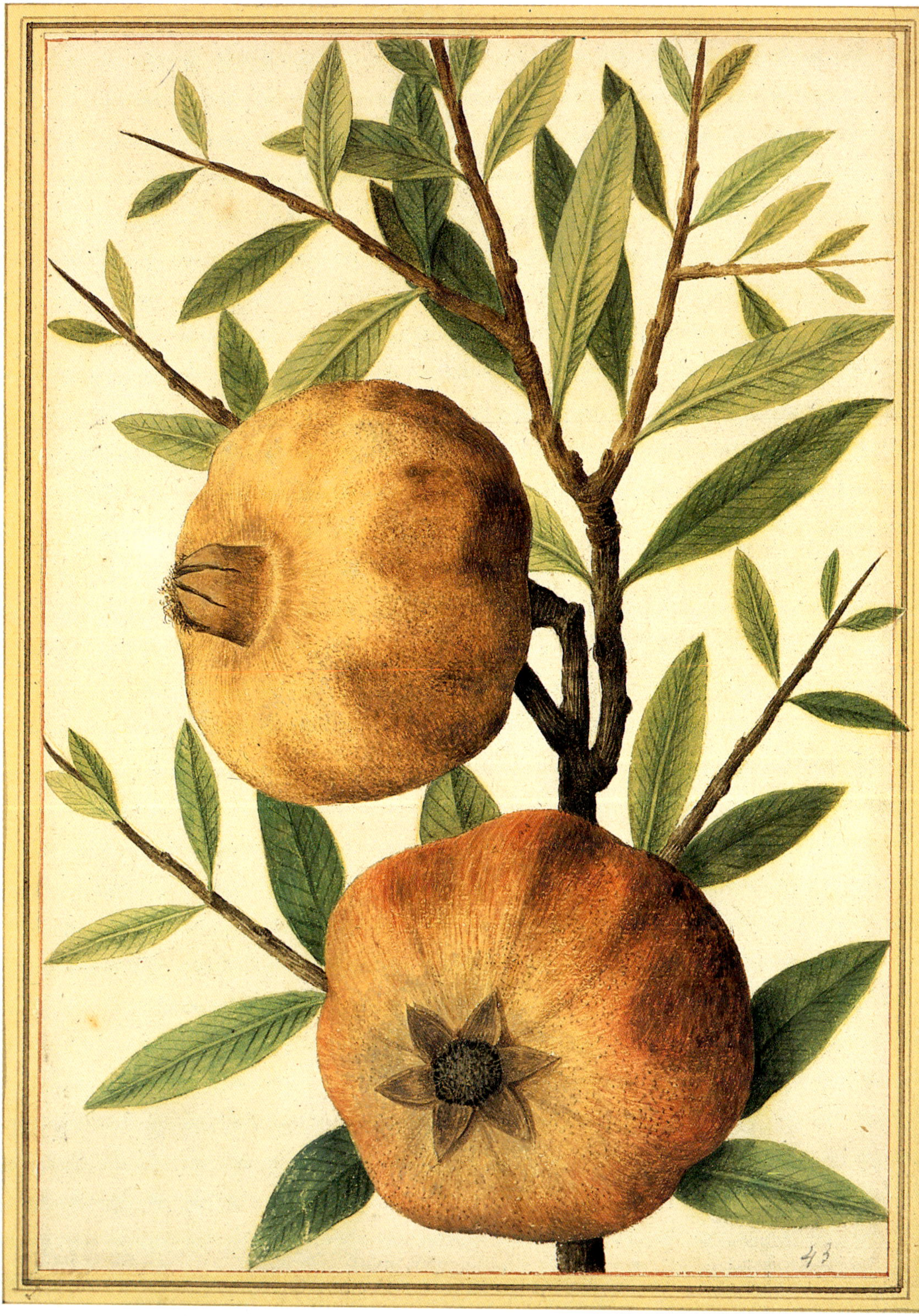
43

September

10

11

12

13

14

15

16

POMEGRANATE
Pomegranate, *Punica granatum* (L.)
Watercolours and bodycolours within a ruled red ink border, extending beyond it on the right (2 and 3 mm).
21.6 x 14.7 cm (8 3/8 x 5 3/4 in.) 1962-7-14-1(45)

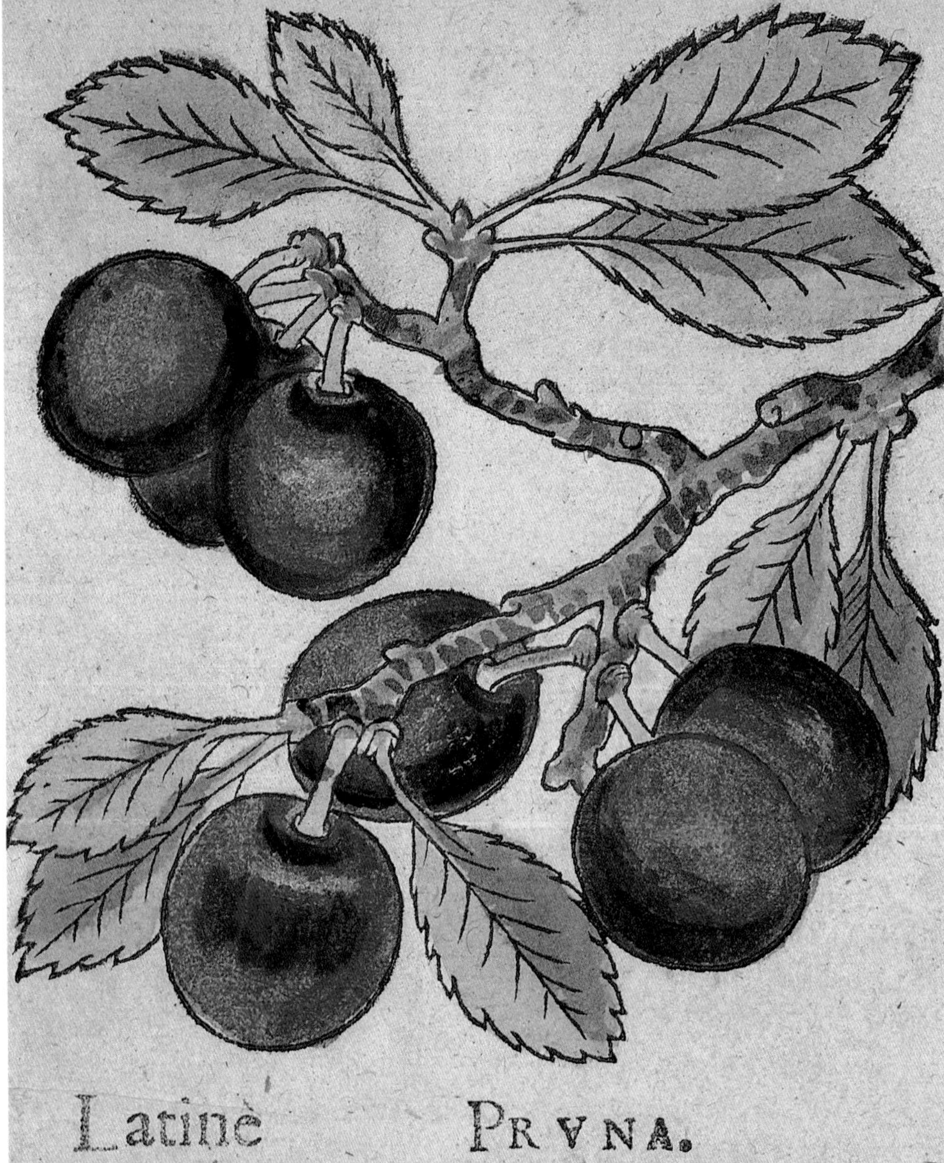

Latinè PRVNA.
Gallicè PRVNES.
Anglicè DAMECONS.

September

17

18

19

20

21

22

23

BULLACE (detail)
La Clef des Champs, 1586
Woodcut, hand-coloured
7.2 x 6.5 cm (2 7/8 x 2 5/8 in.)
Sloane Min. 269

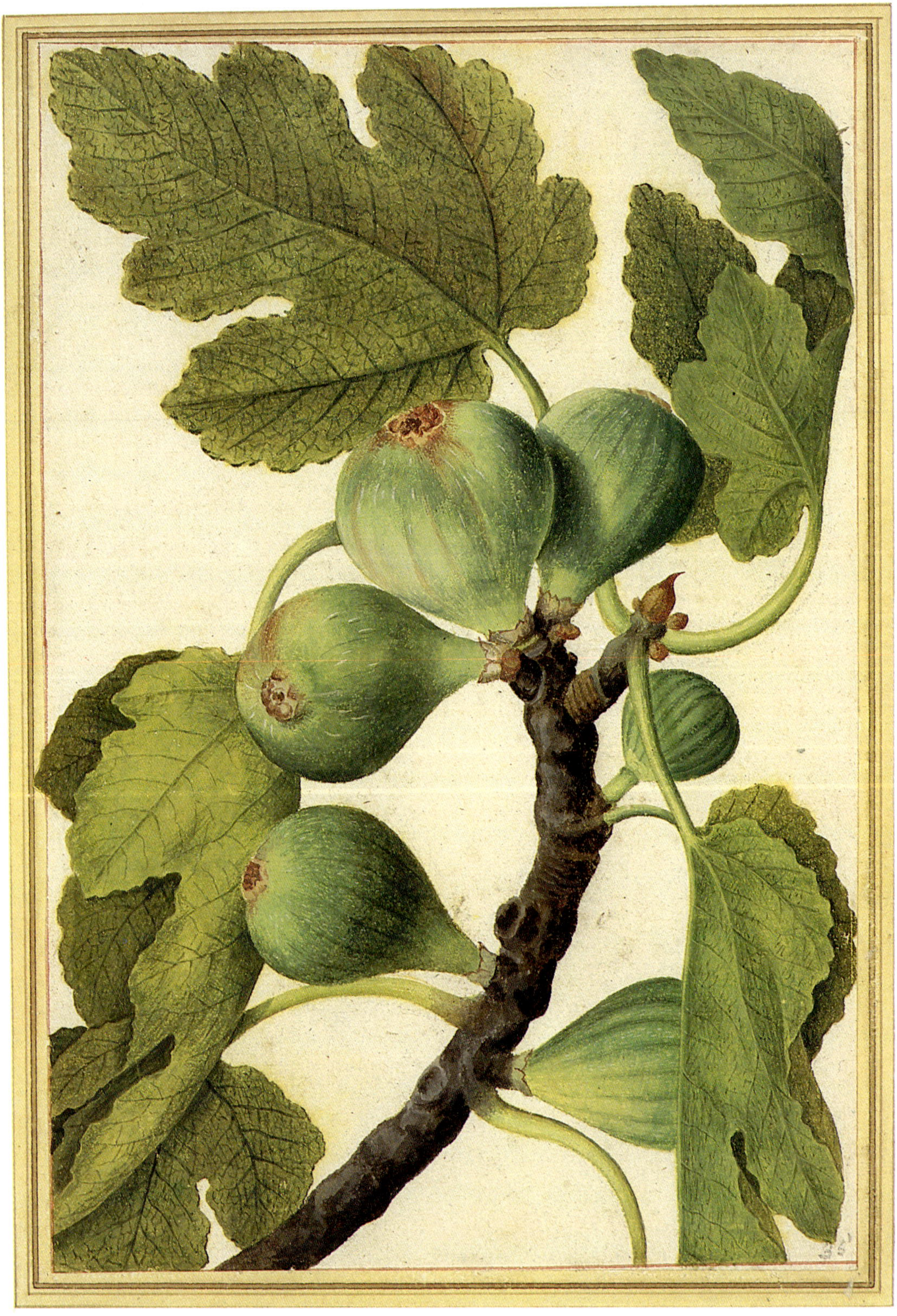

September

24

25

26

27

28

29

30

FIG
Common Fig, *Ficus carica* (L.)
Watercolours and bodycolours within a ruled red ink border, partly cut away on the right.
21.5 x 13.9 cm (8 1/2 x 5 3/8 in.) 1962-7-14-1(37)

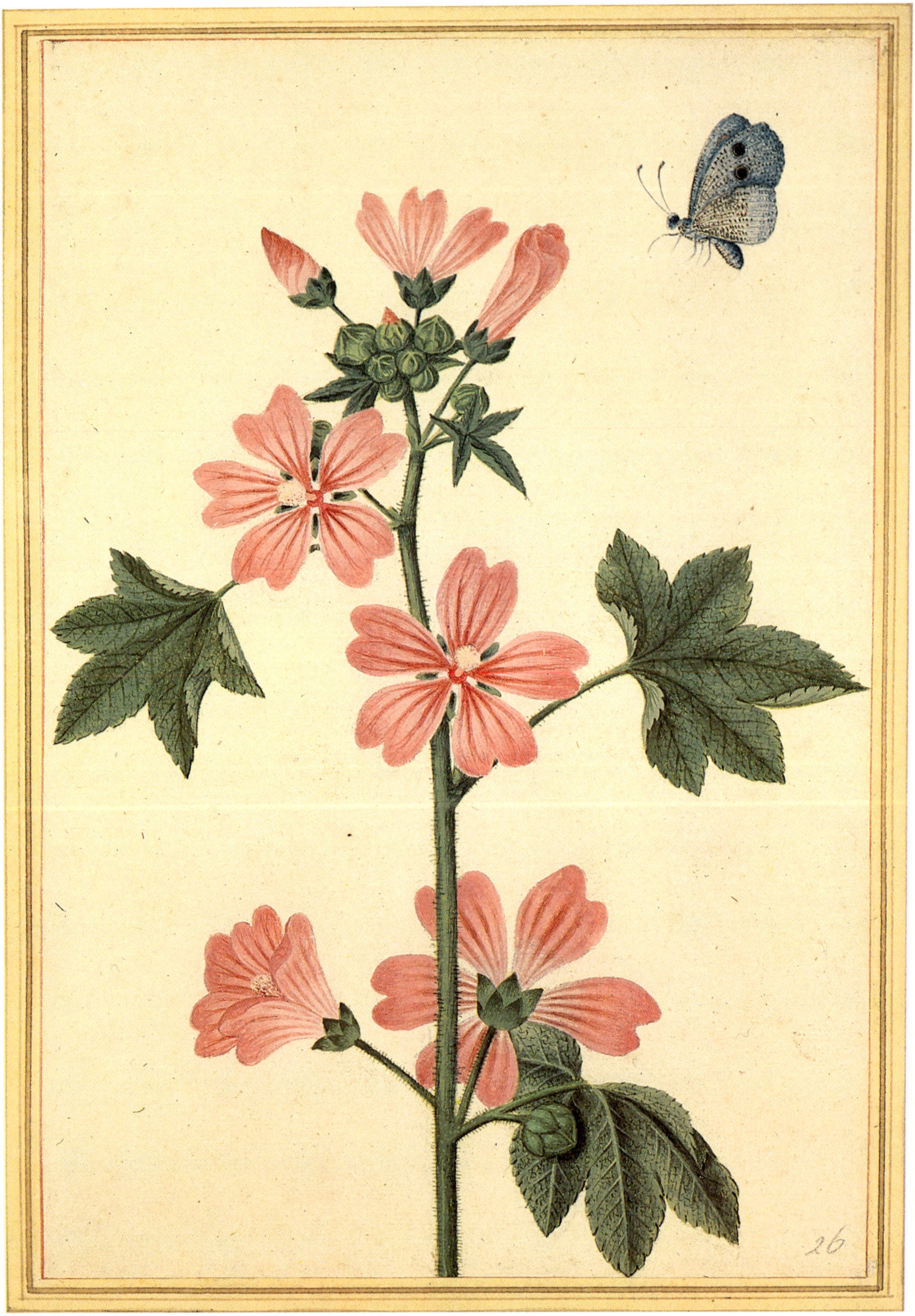
26

October

1

2

3

4

5

6

7

MALLOW AND BUTTERFLY
Common Mallow, *Malva sylvestris* (L.); the butterfly is not identifiable.
Watercolours and bodycolours within a ruled red ink border, partly cut away at the top and bottom.
21.4 x 14.2 cm (8 3/8 x 5 5/8 in.) 1962-7-14-1(28)

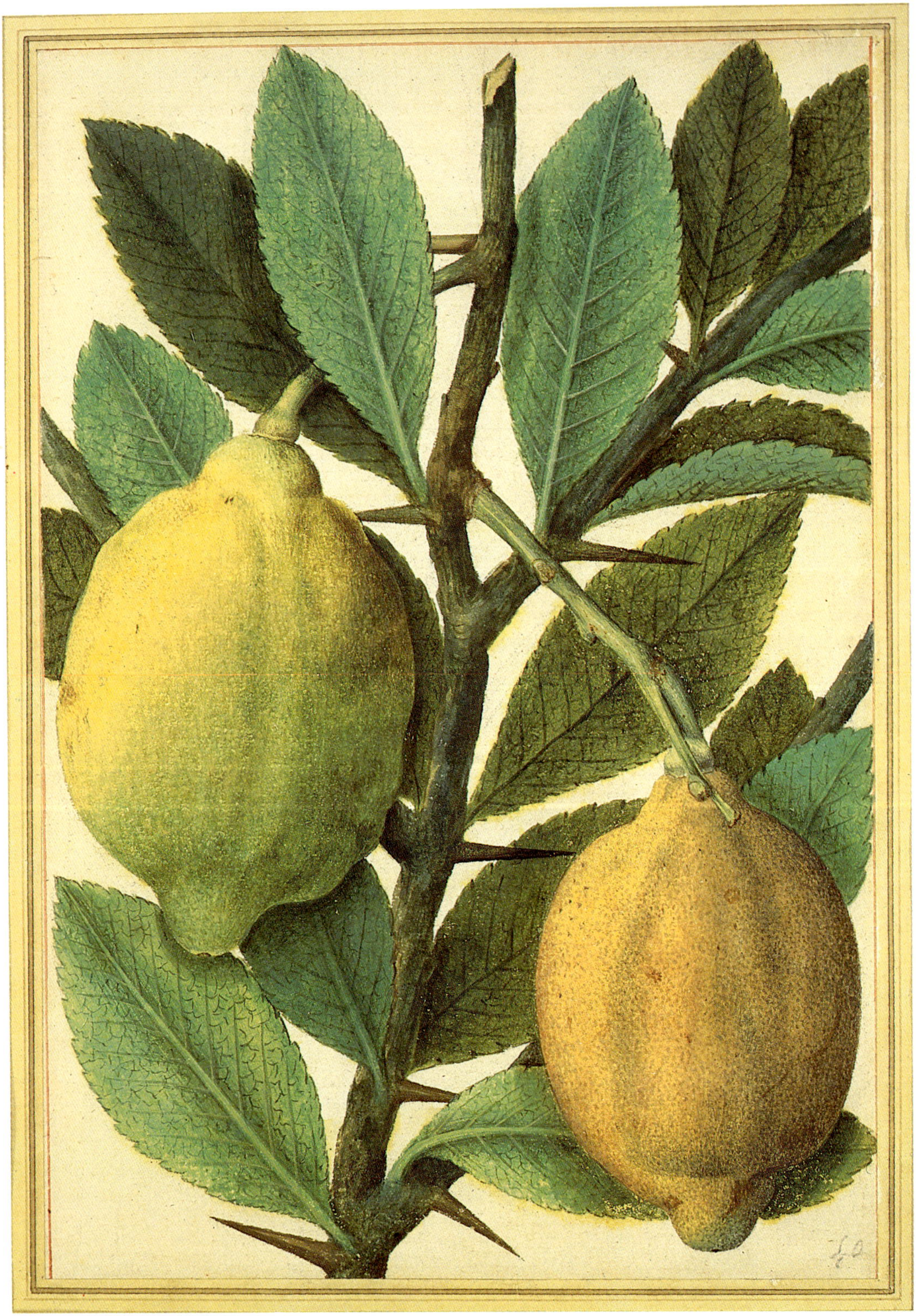

October

8

9

10

11

12

13

14

LEMON
Lemon, *Citrus limon* (L.) N. L. Burman
Watercolours and bodycolours within a ruled red ink border, largely cut away at the bottom.
21.8 x 14.6 cm (8 1/2 x 5 3/4 in.) 1962-7-14-1(42)

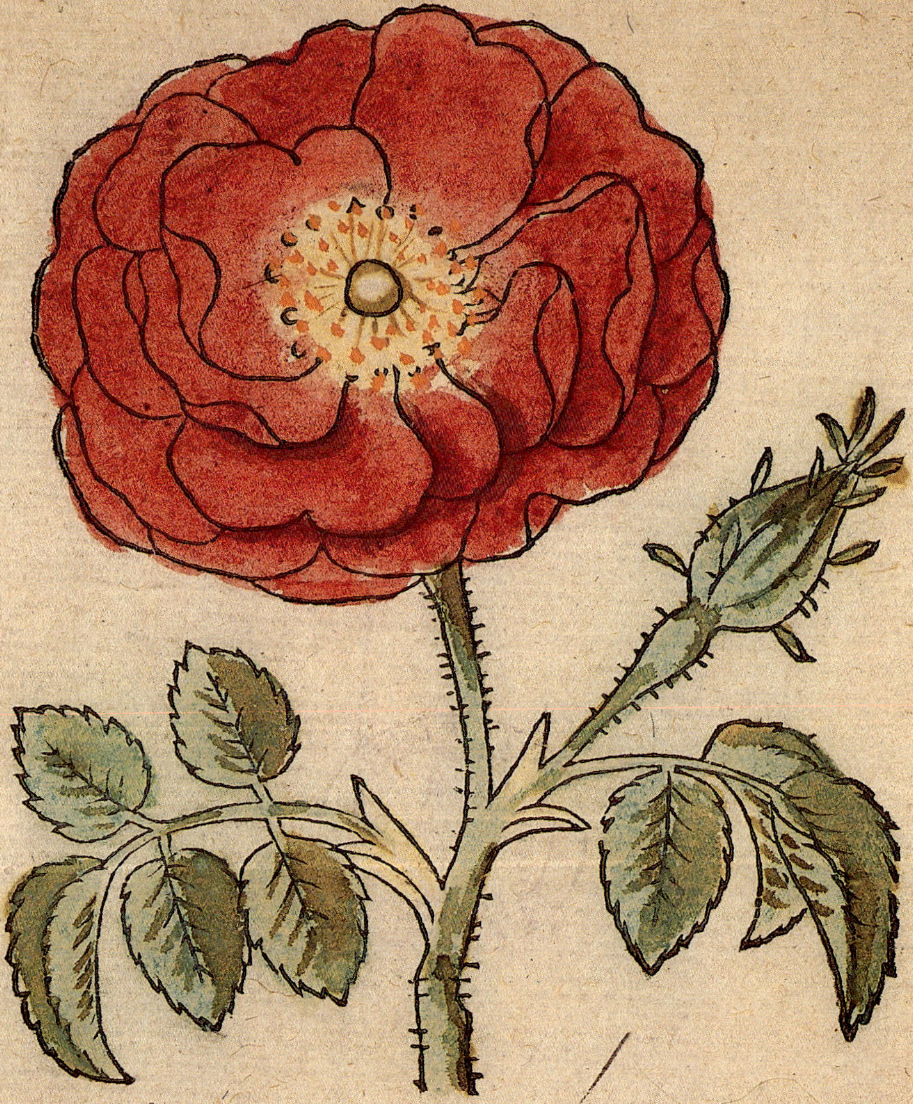

Latinè ROSA RVBRVM.
Gallicè ROSE ROVGE.
Anglicé RED ROSE.

October

15

16

17

18

19

20

21

FRENCH ROSE (detail)
La Clef des Champs, 1586
Woodcut, hand-coloured
7.7 x 6.2 cm (3 x 2 3/8 in.)
Sloane Min. 269

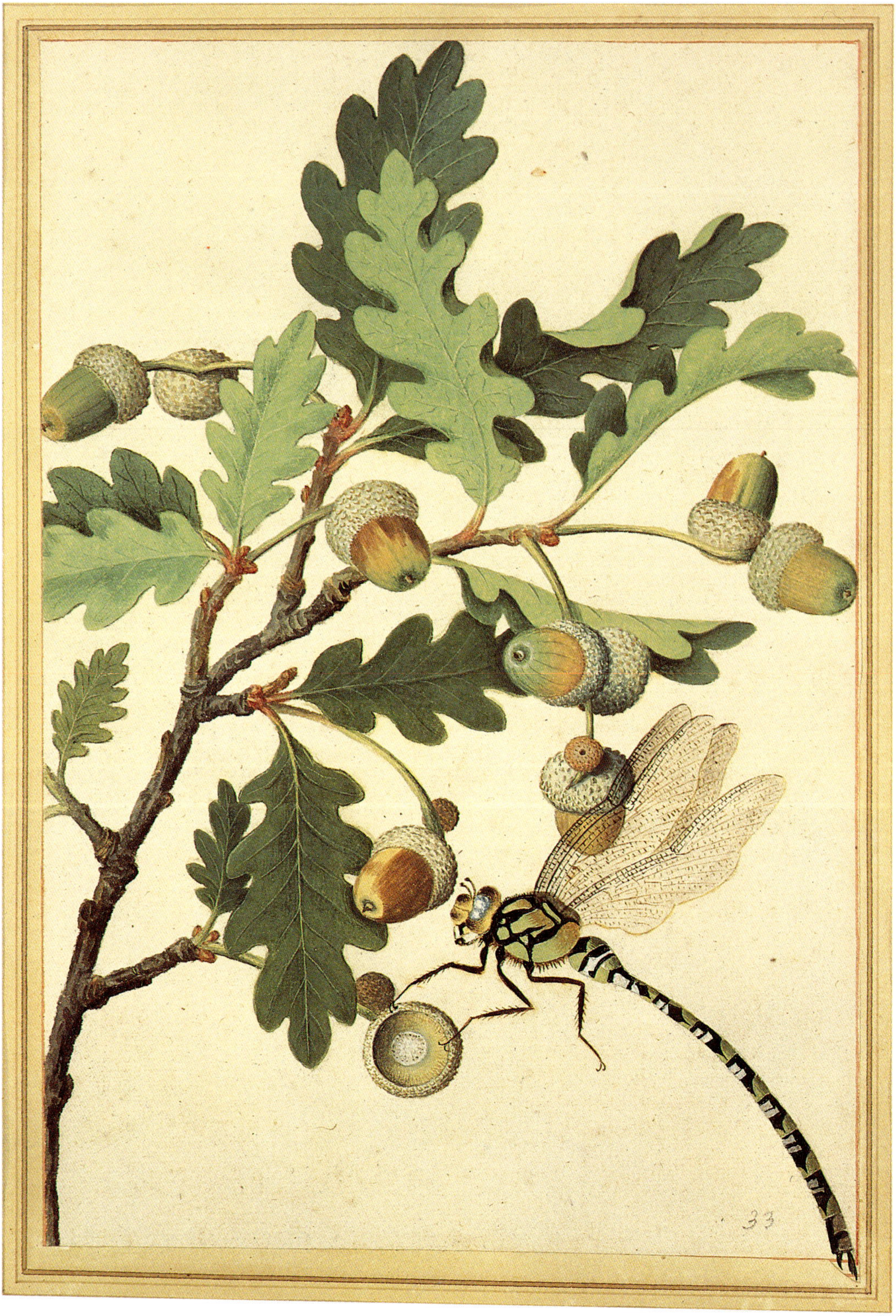
33

October

22

23

24

25

26

27

28

Oak and Dragonfly
Pedunculate Oak, *Quercus robur* (L.); Dragonfly, *Aeshna cyanea* (Müller)
Watercolours and bodycolours with slight oxidisation on the dragonfly; within a ruled red ink border, beyond which the tail of the insect extends (3 mm).
21.3 x 14.2 cm (8 3/8 x 5 5/8 in.) 1962-7-14-1(35)

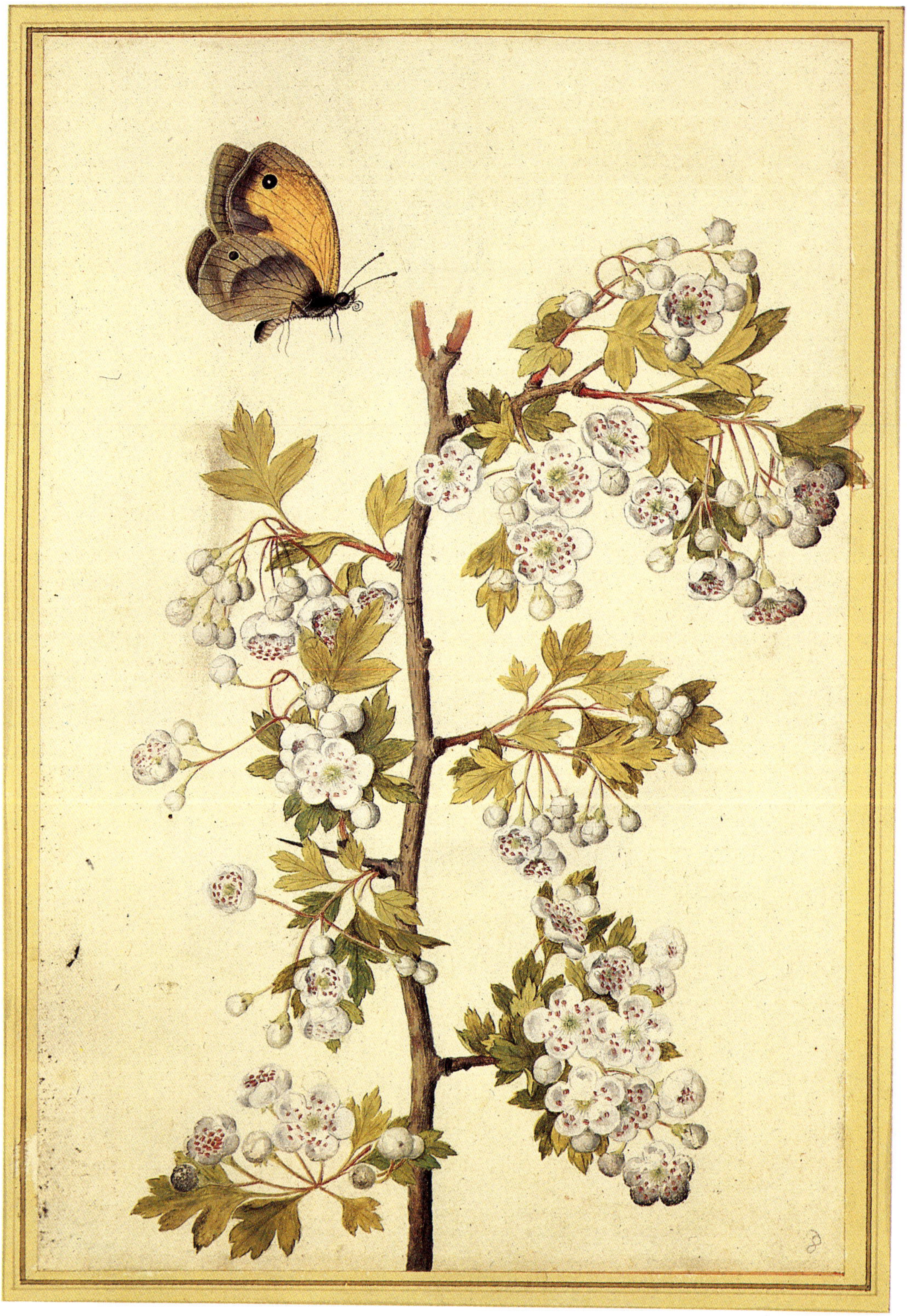

October/November

29

30

31

1

2

3

4

MAYFLOWER AND MEADOW BROWN BUTTERFLY
Hawthorn (Mayflower), *Crataegus monogyna* Jacquin; Meadow Brown Butterfly, *Maniola jurtina* (L.)
Watercolours and bodycolours with some oxidization nearest the borders, within a ruled red ink border, partly cut away on all sides.
21.6 x 14 cm (8½ x 5½ in.) 1962-7-14-1(10)

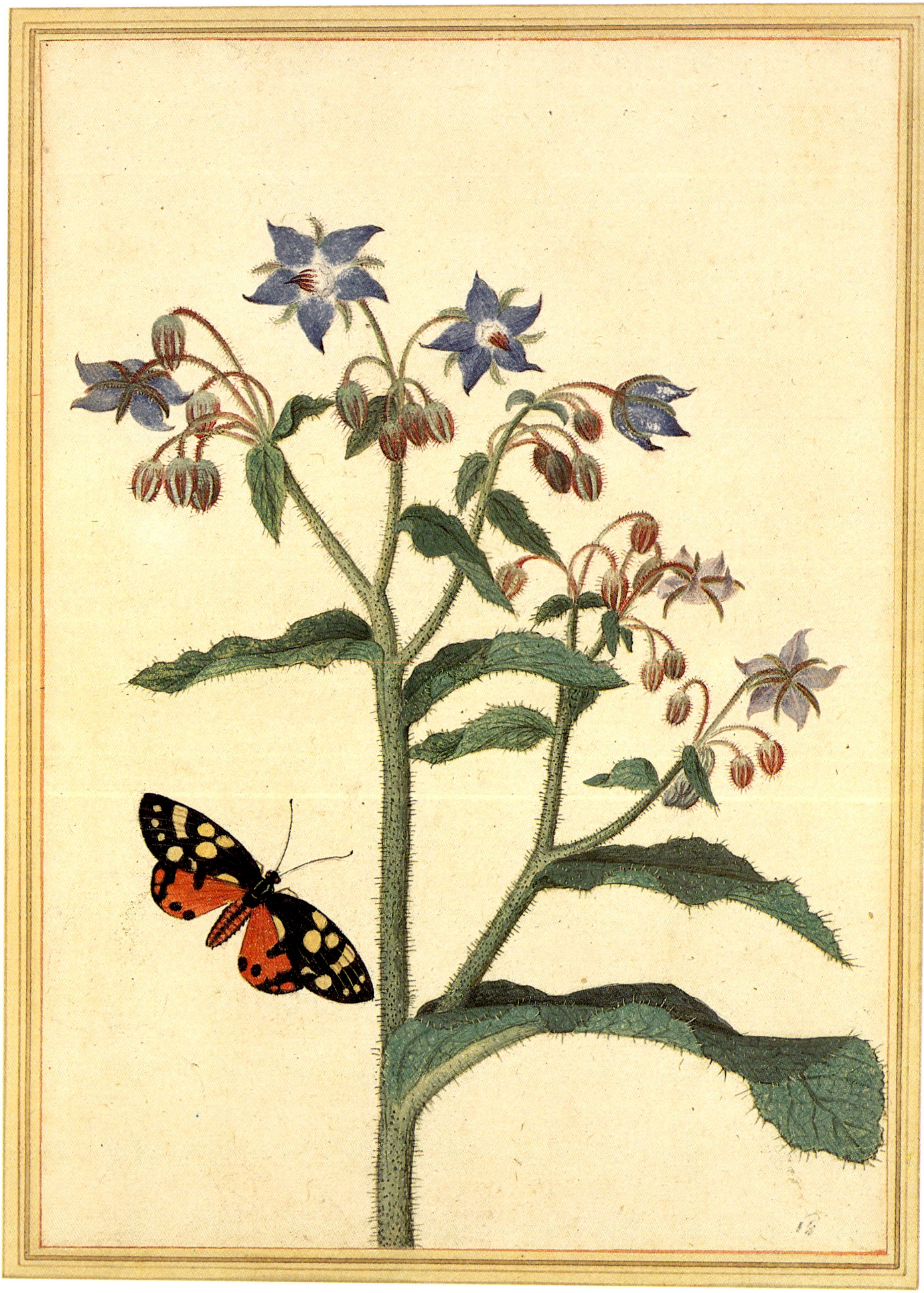

November

5

6

7

8

9

10

11

BORAGE AND SCARLET TIGER MOTH
Common Borage, *Borago officinalis* (L.); Scarlet Tiger Moth, *Callimorpha dominula* (L.)
Watercolours and bodycolours slightly flaking on one of the flowers; within a ruled red ink border, partly cut away at the right.
21.2 x 14.7 cm (8 3/8 x 5 1/4 in.) 1962-7-14-1(20)

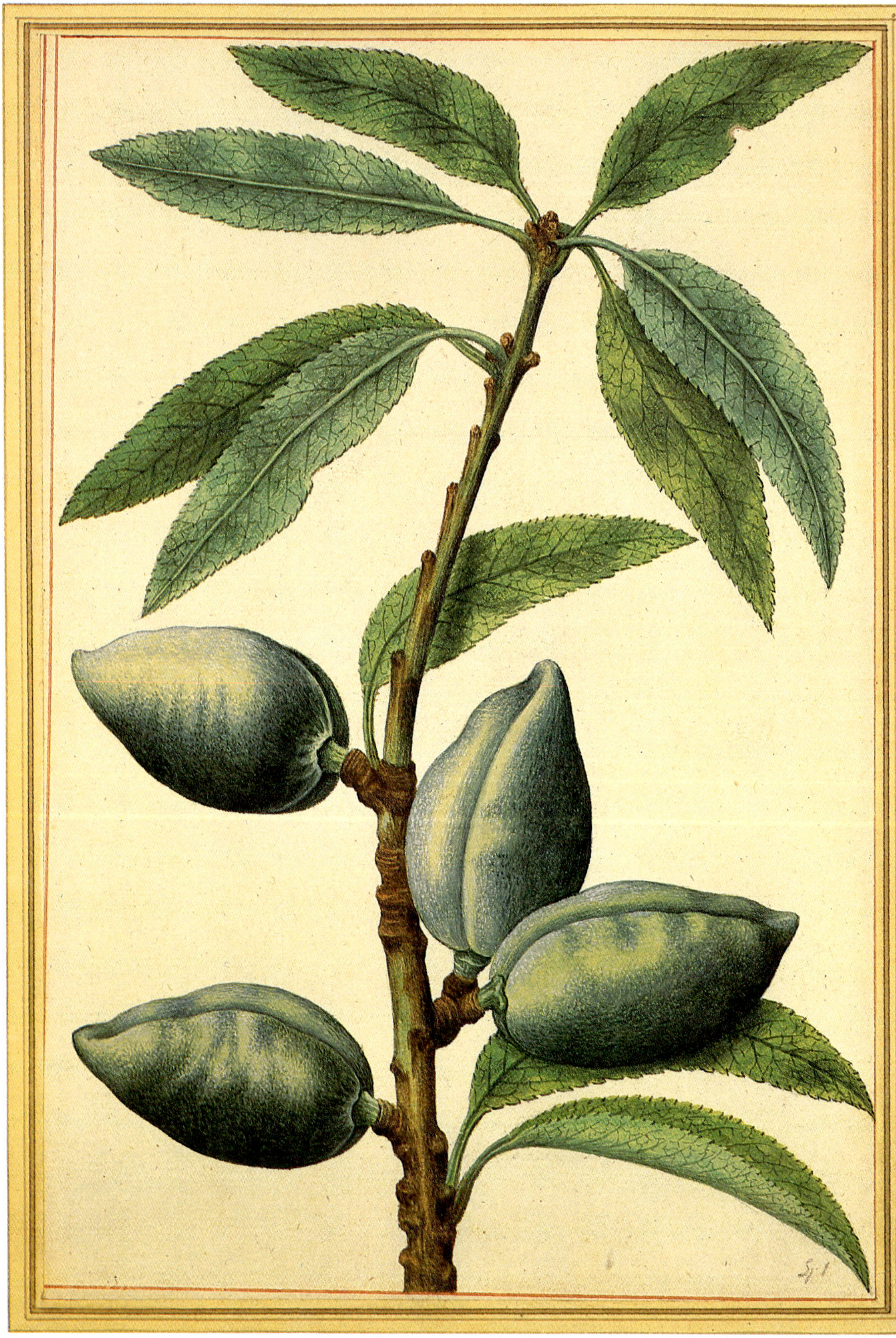

November

12

13

14

15

16

17

18

ALMOND
Almond, *Prunus dulcis* (Miller) D. A. Webb
Watercolours and bodycolours within a ruled red ink border.
21.7 x 14.4 cm (8 1/2 x 5 5/8 in.) 1962-7-14-1(43)

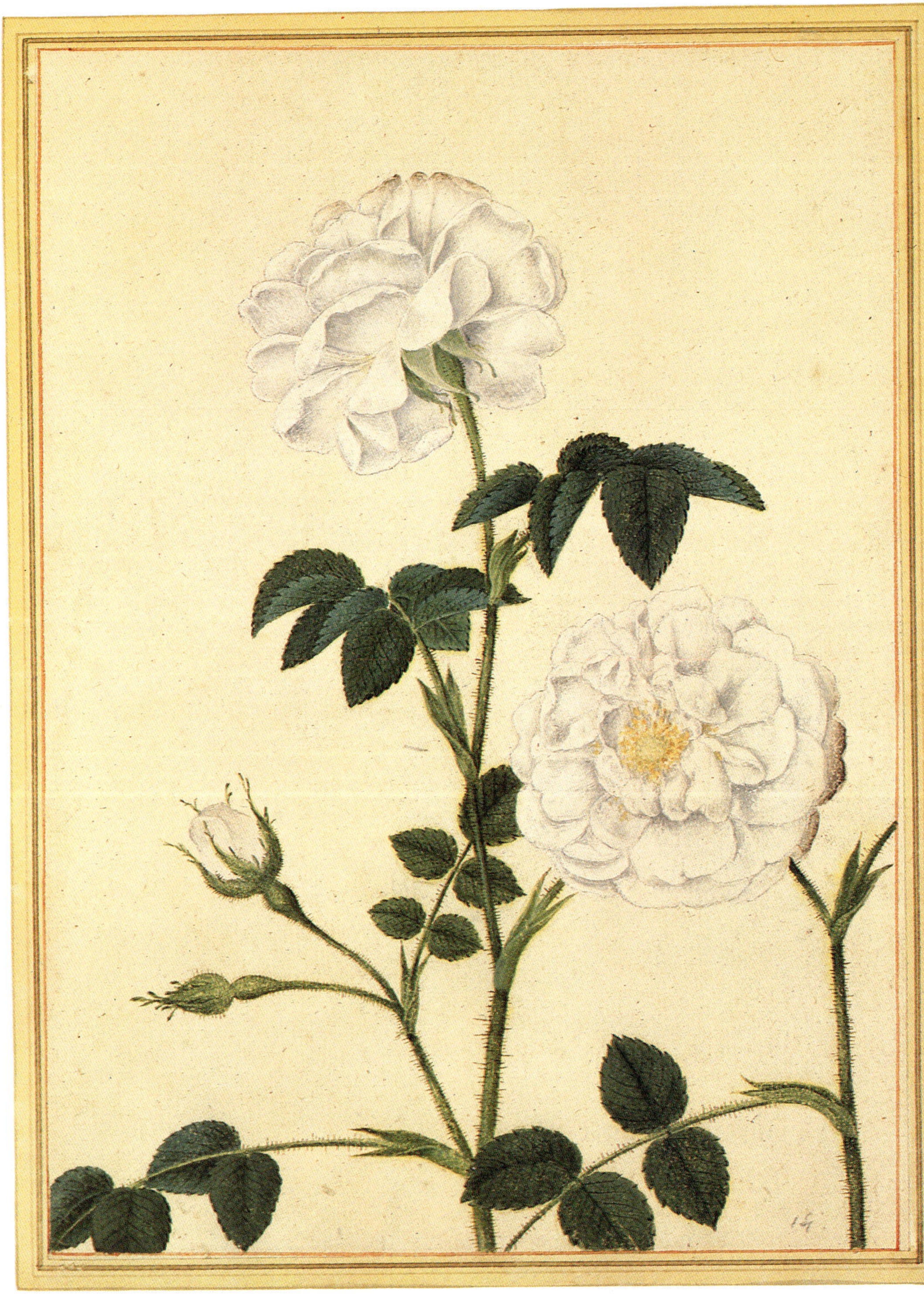
14

November

19

20

21

22

23

24

25

ALBA ROSE
Alba Rose, *Rosa alba* (L.)
Watercolours and bodycolours slightly oxidized at the top and right, within a ruled red ink border.
21 x 14.9 cm ($8\frac{1}{4}$ x $5\frac{7}{8}$ in.) 1962-7-14-1(16)

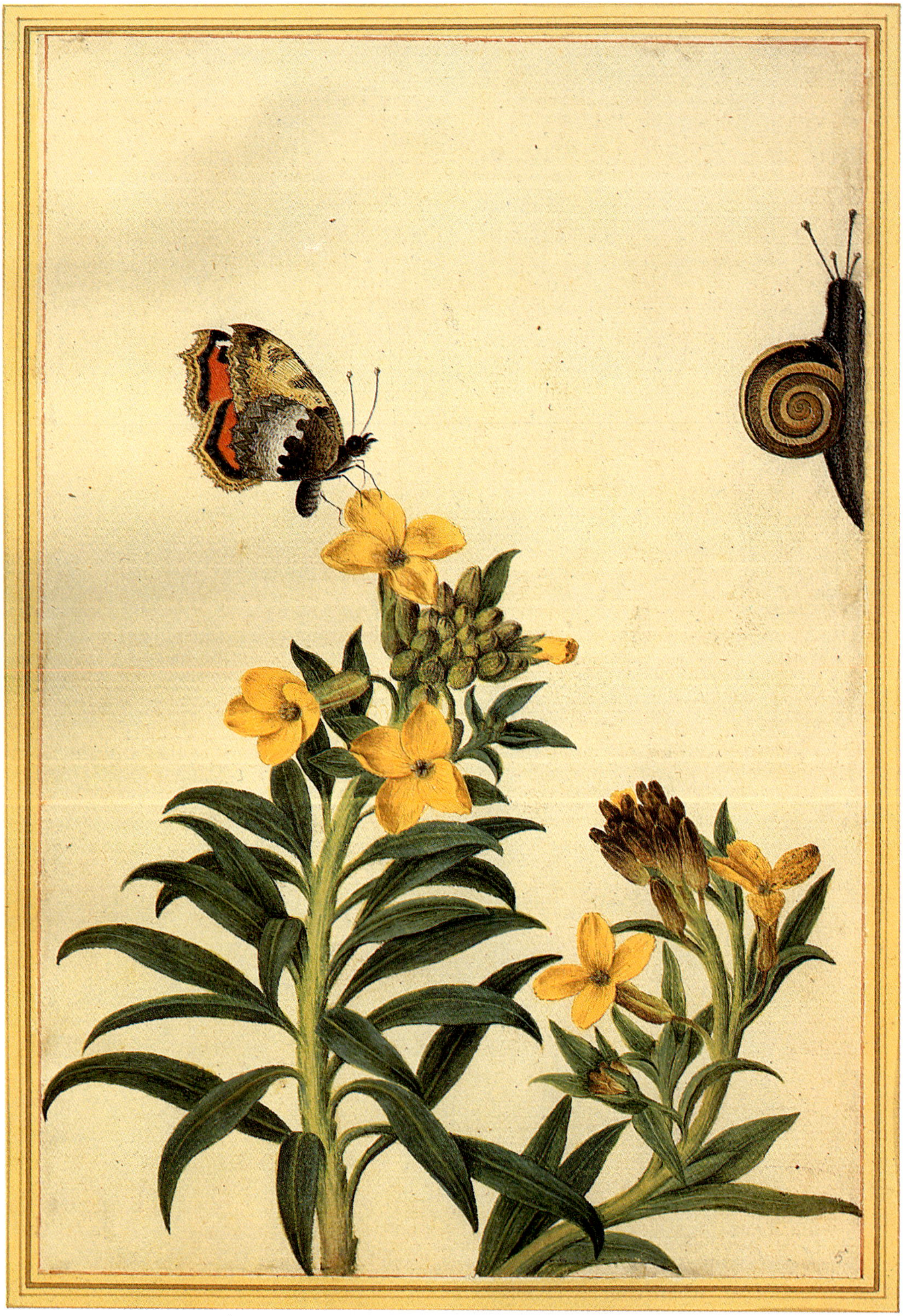
5

November/December

26

27

28

29

30

1

2

WALLFLOWER, SMALL TORTOISESHELL BUTTERFLY, AND SNAIL
Wallflower, *Cheiranthus cheiri* (L.); Small Tortoiseshell Butterfly, *Aglais urticae* (L.); Snail, *Cepaea nemoralis* (L.)
Watercolours and bodycolours with slight flaking near the left wing tip; some oxidization on the flowers; within a ruled red ink border.
21.4 x 14.2 cm (8½ x 5⅝ in.) 1962-7-14-1(7)

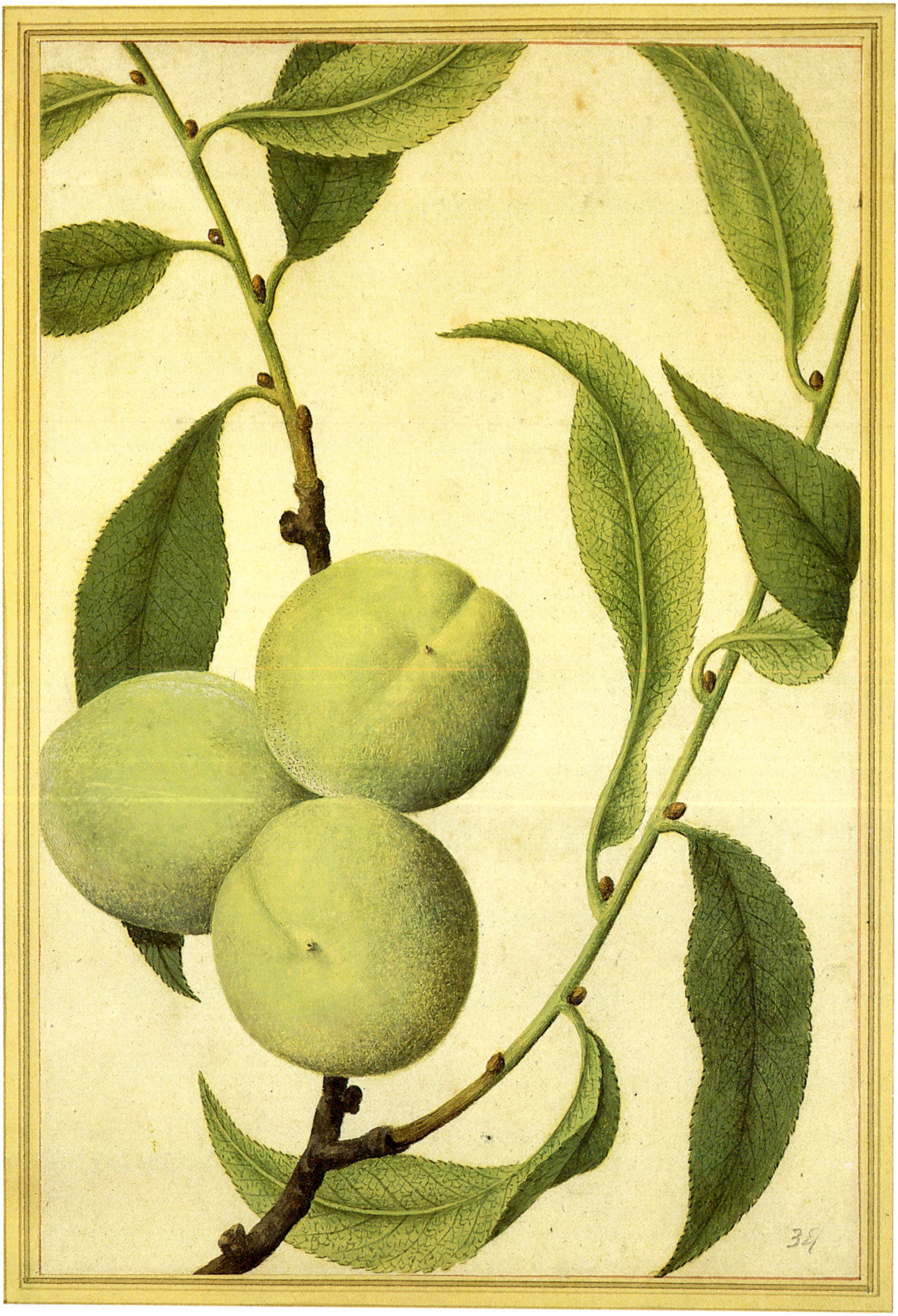

December

3

4

5

6

7

8

9

PEACH
Peach, *Prunus persica* (L.) Batsch
Watercolours and bodycolours within a ruled red ink border.
21.3 x 14 cm (8 3/8 x 5 1/2 in.) 1962-7-14-1(36)

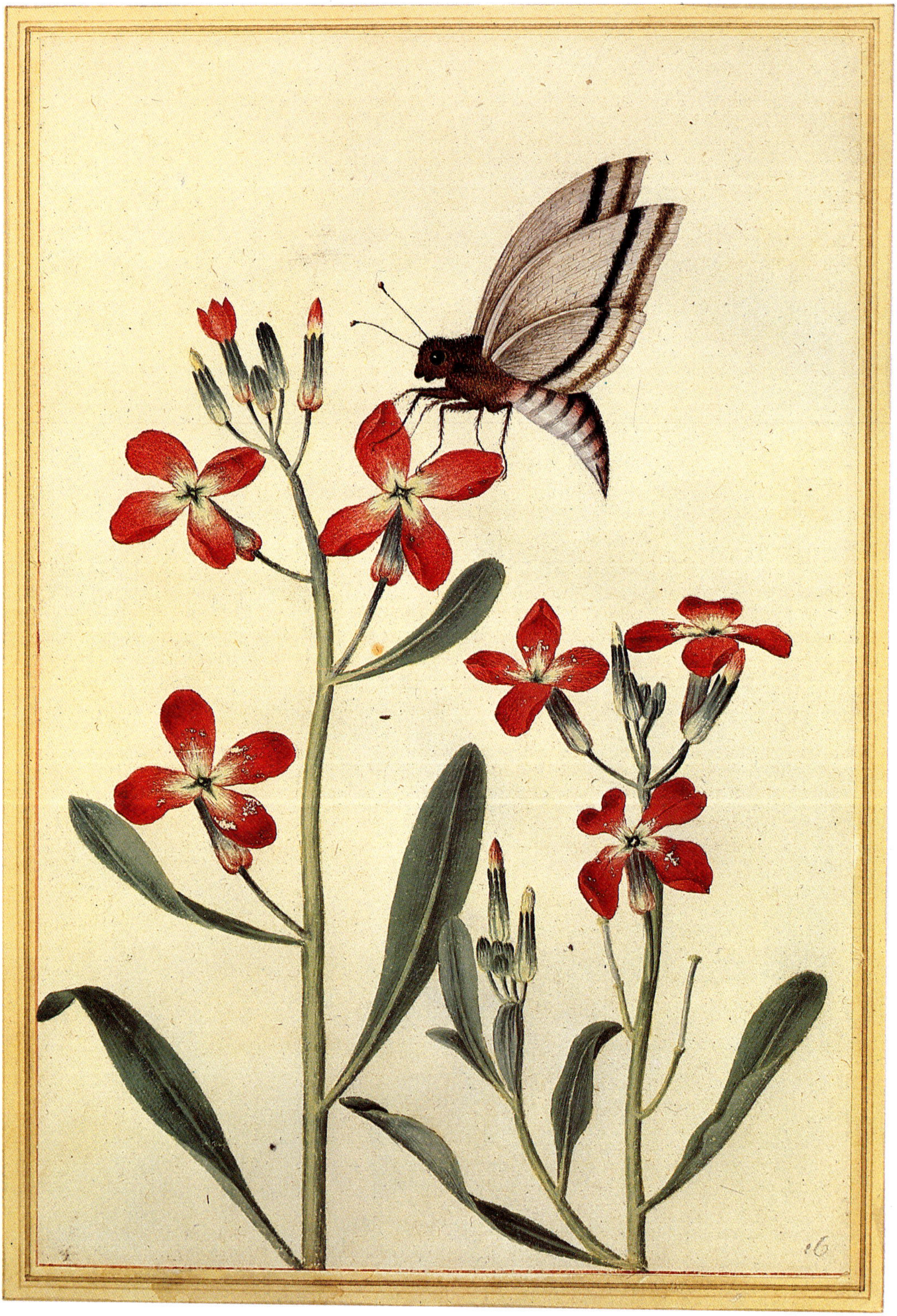

December

10

11

12

13

14

15

16

GILLIFLOWER AND PRIVET HAWK MOTH
Stock (Gilliflower), *Matthiola incana* (L.) R. Brown; Privet Hawk Moth, *Sphinx ligustri* (L.)
Watercolours and bodycolours with slight flaking on the flowers; within a ruled red ink border, cut away on the top and right.
21.5 x 14 cm (8½ x 5½ in.) 1962-7-14-1(18)

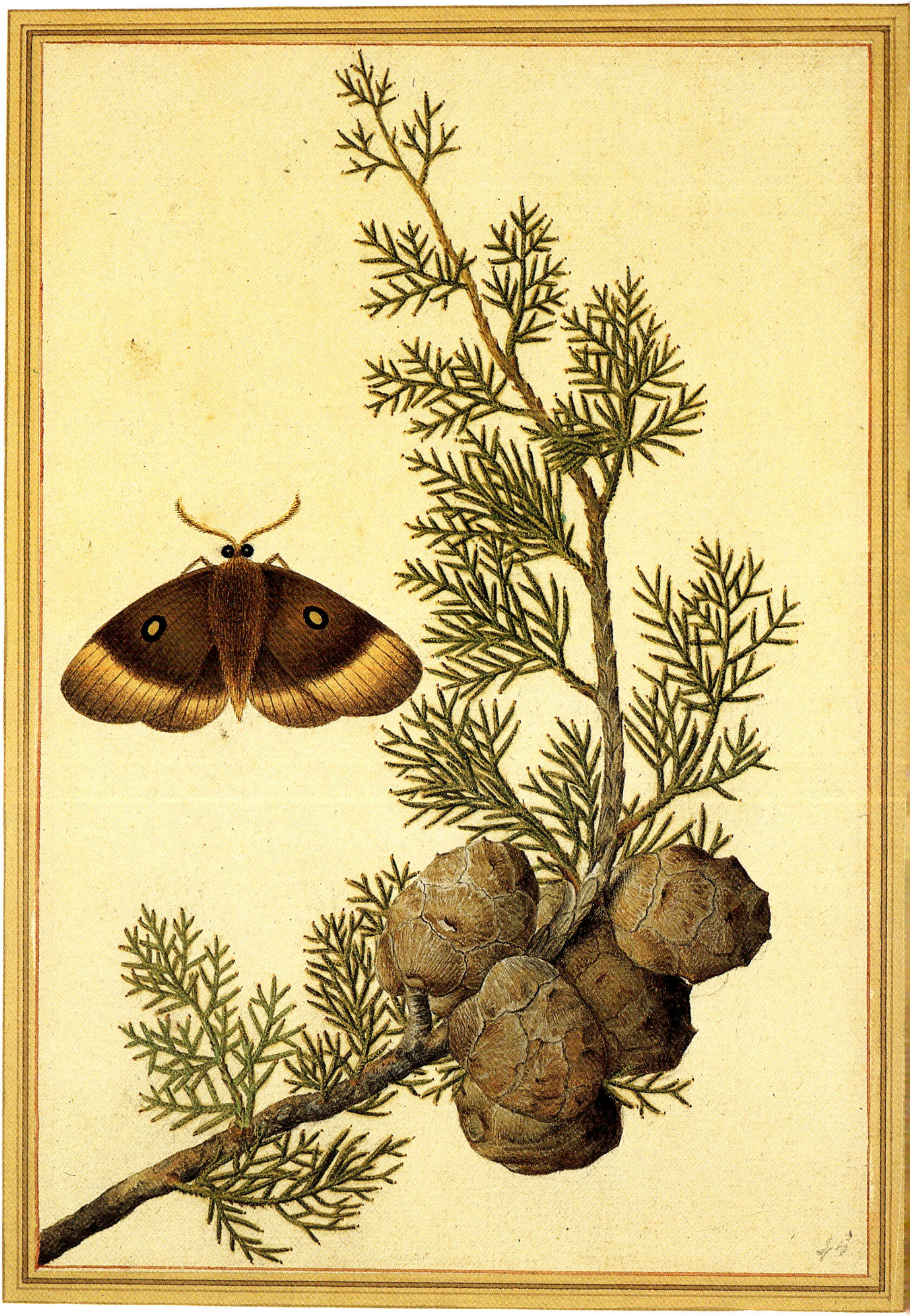

December

17

18

19

20

21

22

23

CYPRESS AND MOTH
Cypress, *Cupressus sempervirens* (L.); Oak Eggar Moth, *Lasiocampa quercus* (L.)
Watercolours and bodycolours within a ruled red ink border.
21.2 x 14.2 cm (8 3/8 x 5 5/8 in.) 1962-7-14-1(47)

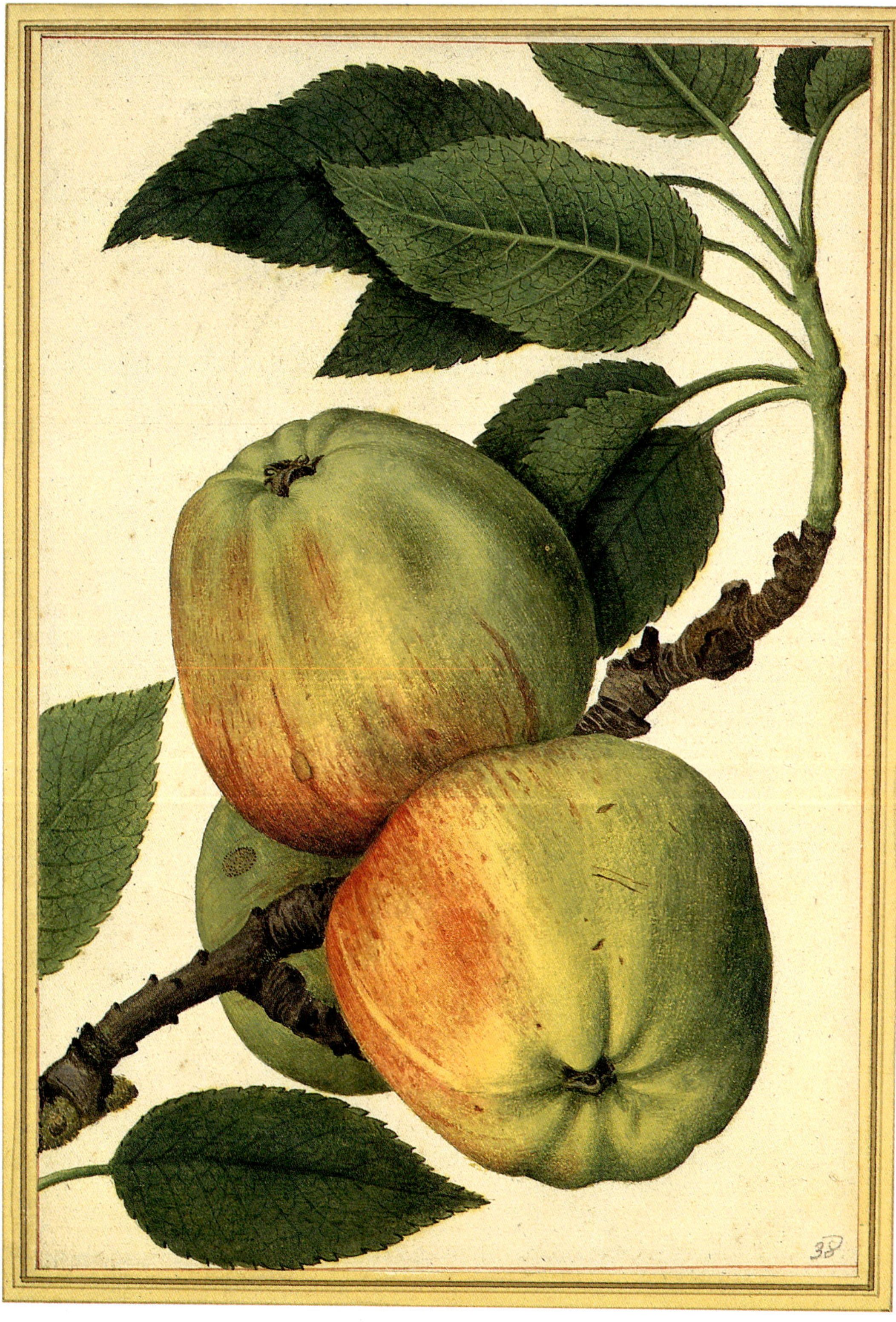
38

December

24

25

26

27

28

29

30 31

APPLE
Apple, *Malus pumila* Miller
Watercolours and bodycolours within a ruled red ink border.
21.5 x 14.1 cm (8 1/2 x 5 1/2 in.) 1962-7-14-1(40)